Sanborn Tenney

Geology

For Teachers, Classes, and Private Students

Sanborn Tenney

Geology
For Teachers, Classes, and Private Students

ISBN/EAN: 9783337176785

Printed in Europe, USA, Canada, Australia, Japan

Cover: Foto ©Paul-Georg Meister /pixelio.de

More available books at **www.hansebooks.com**

THIS VOLUME

IS

RESPECTFULLY DEDICATED

TO

𝔗𝔢𝔞𝔠𝔥𝔢𝔯𝔰,

BY

THEIR FRIEND AND FELLOW-LABORER,

THE AUTHOR.

PREFACE.

THIS volume is designed to present, in a clear manner, the leading facts and principles of Geology. As its title indicates, it is not only intended to prepare the Teacher to give oral instruction upon this important subject, but is especially adapted for a Text-Book in our Common and High Schools, Normal Schools, Academies, and other Seminaries: it is believed that it will also be useful to the general reader.

It will be found wider in its range than most works bearing the title *Geology;* inasmuch as it describes all the common Minerals, and puts the reader in possession of the most important facts about the Vegetable and the Animal Kingdom,—yet all tributary to the general subject.

The book contains within itself all that is necessary to make it entirely intelligible. It does not presuppose a knowledge of any of the subjects of which it treats.

The facts have been drawn from the whole domain of science. Appropriate reference and due acknowledgment, however, are made in the body of the work, to those eminent authors whose writings have been especially consulted.

A large number of the wood-cut illustrations have been drawn from Nature; the others have been copied mainly from the Geological Reports of the different States. With few exceptions, all the subjects represented are common in this country.

In the preparation of this volume, the author has received great encouragement, both from his personal friends, and from teachers generally; for which he now returns his sincere thanks.

He wishes to express his particular obligations to Professors Agassiz and Guyot, for the kind interest they have manifested in this undertaking, and for the great aid they have given by their valuable suggestions. Nor would he forget the favors received from Dr. D. F. Weinland, H. James Clark, Esq., and F. W. Putnam, Esq.

He desires also to return his thanks to the Hon. George S. Boutwell, Professor William Russell, George B. Emerson, LL.D., Professors Alpheus Crosby, Marshall Conant, Hermann Krüsi, the Rev. B. G. Northrop, Thomas D. Adams, Esq., Professors W. F. Phelps, and A. J. Robinson,—from all of whom he has received special encouragement, and important suggestions.

He would be wanting in proper gratitude, if he failed to acknowledge his obligations to the Philadelphia Academy of Natural Sciences, and especially to the accomplished and courteous Librarian, for free access to the Library of that Institution while the sheets were at press.

Nor ought these brief acknowledgments to close without an expression of grateful remembrance to the Rev. J. W. Guernsey, who, many years ago, first turned the attention of the author to those great subjects which he has here endeavored to discuss.

It is but justice to state that the plan of this volume was made, and the work commenced five years ago; that more than a year ago it was considered ready for the press, and notice given to that effect. But unavoidable circumstances have prevented its issue till now. Although in the meantime every page has been carefully revised, the plan, which, so far as the writer is informed, is different from that of any other work, remains the same as it was at the outset.

That this volume may tend to promote the study of Nature, in the right spirit, and thereby advance the cause of popular education, is the sincere desire of the author.

S. T.

OCTOBER, 1859.

CONTENTS.

CHAPTER I.
GENERAL STATEMENT OF THE SUBJECT 13

CHAPTER II.
THE EARTH CONSIDERED AS A PLANET 16

CHAPTER III.
CHEMICAL CONSTITUTION OF THE EARTH 20

CHAPTER IV.
MINERAL CONSTITUTION OF THE EARTH 26

CHAPTER V.
THE ROCKS WHICH COMPOSE THE EARTH 65

Section I.—Two great Classes recognised	65
II.—Description of the Unstratified Rocks	68
III.—Description of the Stratified Rocks	81

CHAPTER VI.

General view of the Vegetable and the Animal Kingdom, preparatory to the Study of the Remains of Plants and Animals in the Rocks 91

Section I.—The Vegetable Kingdom	91
II.—The Animal Kingdom	102

CHAPTER VII.

Fossils, and Classification of the Rock Formations . . . 132

Section I.—Fossils	132
II.—Classification of the Rock Formations	139

CHAPTER VIII.

Brief description of the several Systems of Fossiliferous Rocks 143

Section I.—Silurian System	143
II.—Old Red Sandstone System	150
III.—Carboniferous System	155
IV.—New Red Sandstone System	169
V.—Oölitic System	175
VI.—Cretaceous System	181
VII.—Tertiary System	185
VIII.—Drift	192
IX.—Alluvium	202

CHAPTER IX.

Geological Changes now going on, and the Agencies by which they are produced 216

    Section I.—Aqueous Agencies 217
           II.—Igneous Agencies 239
          III.—Organic Agencies 258

CHAPTER X.

Concluding Remarks 268

GLOSSARY 283

INDEX 313

GEOLOGY.

CHAPTER I.

GENERAL STATEMENT OF THE SUBJECT.

Geology is the science which treats of the earth. Everything which relates to the physical condition of our planet, may come within its domain. Its special province, however, is the investigation of the rock formations of the earth, their constitution and arrangement, and the changes they are undergoing, and have undergone in past ages.

Geology does not attempt to explain the origin of the earth, but, recognising it as the handiwork of a Divine Creator, seeks only to investigate its present condition and past history.

The facts of this science are obtained by a careful examination of everything which forms a part of the accessible matter of the globe,—loam, clay, sand, gravel, and boulders, as well as the solid rocks, and their organic contents.

Chemistry treats of the elements of matter, their properties, and laws of combination.

Mineralogy describes the separate minerals which compose the earth.

Botany is the scientific examination of the various forms of vegetable life.

Zoölogy treats of the animal kingdom.

In geological investigations it is necessary to refer to all these branches of science.

Until a comparatively recent period, it was a very common impression, that the rocky strata were created in their present form and position; but geology teaches that they have undergone many changes, the proofs of which exist in the rocks themselves. We find abundant evidence that the physical condition of the earth has been widely different at different epochs, and that many successive races of animals and plants have flourished on its surface, differing from one another, and from those now living.

Geology not only teaches much that is instructive and interesting, concerning the past history of the earth, but it also gives important information in regard to the rocks and minerals of practical use to man; as the experienced geologist can decide, with a good degree of accuracy, whether certain substances may, or may not, be found in a given region.

As an educational branch, Geology especially claims our attention. Its study is calculated to enlarge and ennoble all the faculties of the mind. Nature and Revelation are man's great teachers — both are indispensable for his perfect mental and moral development.

Therefore, whether we study geology for its practical advantages, or for the higher purpose of becoming acquainted with the great plan which the Author of Nature has made, and carried out, in regard to our earth, the subject becomes one of the greatest importance and highest interest.

CHAPTER II.

THE EARTH CONSIDERED AS A PLANET.

It is an established fact that the earth is an oblate spheroid, the polar diameter being twenty-six miles less than the equatorial; which gives a flattening of thirteen miles at each pole.

If a ball of clay, or any yielding material, be revolved rapidly by means of a wire or rod, for an axis, the revolutions will produce a difference between the two diameters of the ball;—the diameter in the direction of the axis will diminish, while that at right angles to the same will increase. This may serve to explain the present figure of the earth, since it needs only to have been in a soft, or yielding state, in order to be affected by revolution, in the same manner as the ball above mentioned.

Other planets besides the earth exhibit a flattening at the poles, and it is greatest upon those which rotate with the greatest velocity. The earth rotates on its axis at the rate of about 1000 miles per hour, and exhibits a difference between its equatorial and polar diameters of a little more than 26 miles; while Jupiter rotates on its axis at the rate of 28,000 miles an hour, and its equatorial diameter exceeds the polar by 6000 miles.

It is estimated that the density of the earth, taken as a whole, is five times as great as that of water, and about two and a half times as great as that of ordinary rocks. In other words, five globes of water, of the same size as the earth, would weigh no more than the earth itself; and two and a half globes of ordinary rock, of the same size as the earth, would only equal the weight of the latter. The density of the earth, then, must increase as we go towards the centre. If this increase be regular, water, at the depth of 362 miles, would be as heavy as quicksilver, and air, at 34 miles, as heavy as water at the surface; and at the centre, rocks would be compressed into one eighth of their ordinary bulk.

Accustomed to regard the surface of the earth as very irregular, unless our ideas in regard to this fact be properly modified, we shall not so readily comprehend the great movements which have taken place in times past. Given irregularities on a globe one inch in diameter, may make it appear exceedingly rough, while the same irregularities would hardly be perceptible on a globe one hundred feet in diameter. In proportion to the size of the two bodies, an orange is much rougher than the earth. Taking the diameter of the earth in round numbers at 8000 miles, and the height of the highest mountain on its surface at 5 miles, then the elevation of the highest mountain compared with the diameter of the globe, is as 1 to 1600; or the thickness of one sheet of paper holds the same relation to the thickness of sixteen hundred sheets, as the highest mountain to the diameter of the globe. Keeping these facts in mind, the student will better understand the elevations and depressions of continents, the uplifting of mountain ranges, and all the movements to which the earth is subject.

The temperature of the surface of the earth depends upon a variety of circumstances; differing in different places, according to distance from the equator, elevation, situation in respect to the ocean, oceanic currents, and prevailing winds, and according to the radiating power of the soil.

The sun does not affect the temperature of the earth below the depth of 100 feet. But for every 45 or 50 feet of descent below that point, the temperature rises about one degree Fahrenheit. If the increase go on at that rate, a point would soon be reached where the heat is sufficient to melt all known substances.

It is believed by most geologists, that the interior of the earth is in a molten state; that the whole earth has been in that condition; and that the crust, or solid portion, is not more than 50 or 100 miles thick. But this theory will be further considered in a subsequent chapter. At present, the rock formations, which may come under actual observation, will afford the beginner in geological investigations, ample field for study. It may be remarked here, however, that, in his examinations of the solid rocks, the student will constantly find the effects of heat.

CHAPTER III.

CHEMICAL CONSTITUTION OF THE EARTH.

CHEMISTRY informs us that the whole matter of our globe is composed of 62 or 65 elements. The following list, divided into two principal groups, and the first into three subordinate ones, contains the names of 36 of those most abundant, and indicates their condition when uncombined, and at ordinary temperatures.

NON-METALLIC ELEMENTS: *Gases*—Oxygen, Hydrogen, Nitrogen, Chlorine, Fluorine. *Liquid*—Bromine. *Solid*—Iodine, Carbon, Sulphur, Phosphorus, Silicon, Boron.

METALS AND METALLOIDS: Potassium, Sodium, Calcium, Barium, Strontium, Magnesium, Aluminium, Iron, Manganese, Cobalt, Nickel, Copper, Cadmium, Bismuth, Lead, Tin, Zinc, Chromium, Antimony, Arsenic, Mercury, Silver, Gold, Platinum.

Sixteen of the above, in various combinations and proportions, compose the great bulk of the earth. Part of the remainder form rare and valuable minerals, and others are seldom found. The sixteen are as follows:—Oxygen, Hydrogen, Nitrogen, Fluorine, Chlorine, Carbon, Sulphur, Phosphorus, Silicon, Aluminium, Potassium, Sodium, Calcium, Magnesium, Iron, and Manganese. Each of these will now be briefly noticed, and some of the most important of the others will be described in a subsequent chapter.

OXYGEN is the most abundant, and most widely diffused of all the elements. It constitutes from one-third to one-half of the crust of the earth, forming a portion of almost every substance. It forms, by weight, one-fifth of the atmosphere, and eight-ninths of water.

This gas is about one-tenth heavier than common air. One hundred cubic inches weigh $34\frac{1}{3}$ grains. It is tasteless, colorless, and inodorous. It is the great supporter of animal life and combustion.

HYDROGEN forms one-ninth part of water. This gas is the lightest of all known substances, weighing only about one-fourteenth as much as air. One hundred cubic inches weigh 2.14 grains. It is never found free in nature. Hydrogen is not a supporter of combustion, but will itself burn freely. When mixed with certain proportions of oxygen, or atmosphere air, it forms a highly explosive and dangerous compound.

How wonderful that the sparkling water is but a combination of two colorless, invisible gases, one of which will ignite and burn, and the other the great supporter of combustion!

NITROGEN constitutes 80 per cent. of the atmosphere, and occurs in combinations, both in the animal and in the vegetable kingdom. It is neither a supporter of respiration nor combustion. It instantly extinguishes a lighted taper, and destroys animal life; and yet, properly mixed with oxygen, it forms the air we breathe.

CHLORINE is a green gas about as heavy as air. Combined with sodium, it forms chloride of sodium, or common salt. Beds of rock-salt are common in many parts of the world. The most extensive are in Poland, Hungary, Austria, and the countries in that quarter of the world. In the United States, salt is obtained in immense quantities by evaporating the water of salt springs. The springs of New York, located mostly at Syracuse and vicinity, yield annually about 6 millions of bushels, and those of Virginia $3\frac{1}{2}$ millions.

FLUORINE occurs in combination with calcium, forming the mineral known under the name of fluor spar, which will be noticed hereafter. Fluorine, combined with hydrogen, forms hydrofluoric acid, used in etching upon glass.

CARBON, in its pure state, is a solid. It constitutes the principal part of all the varieties of mineral coal, which will be noticed hereafter. Plumbago, commonly but incorrectly called Black Lead, is mainly carbon, containing, besides this element, only 5 or 6 per cent. of iron. Plumbago occurs in foliated masses in the older rocks, such as mica-slate and gneiss. Nelson and Washington, N. H., and Sturbridge, Mass., are among the important localities in the United States. In England, the principal locality is in Cumberland. Plumbago is used for the manufacture of lead pencils, crucibles, and for diminishing the friction of machinery. Bitumen, in all its varieties, is mainly carbon. Charcoal is almost pure carbon, and the Diamond is pure crystallized carbon. The principal diamond regions are India, the Ural Mountains, and Brazil. The diamond has been found in the United States in Virginia, North Carolina, and Georgia.

Carbon constitutes from 25 to 50 per cent. of all vegetable matter. In combination with oxygen, it forms carbonic acid, a gas which instantly extinguishes a lighted taper if immersed in it, and is fatal when inhaled. This gas is formed by respiration, by combustion, and by the decay of all animal and vegetable substances. It is often found in old wells, and in caves. It exists largely in the waters of many springs, especially those of Saratoga. Carbonic acid combines with numerous substances, forming a class of substances called carbonates. Chalk and limestone are carbonates of lime.

SULPHUR is a yellow, brittle solid, occurring both native and combined with other substances. It is found most abundantly in volcanic districts. The island of Sicily furnishes a large part of the native sulphur of commerce. It is widely diffused in a state of combination with the metals, forming the sulphurets of iron, lead, copper, zinc, &c. It comprises half the weight of gypsum, existing in that substance as sulphuric acid, the same being the result of the combination of sulphur and oxygen.

Sulphur occurs in many mineral waters, also in plants, and sparingly in animal tissues. It occurs in eggs, which, as is well known, tarnish silver. The sulphur of the egg readily combines with that metal, thus forming the black sulphuret of silver.

PHOSPHORUS never occurs free in nature, but is quite abundant in combinations. With lime it forms the phosphate of lime, which occurs quite abundantly in the crust of the earth; also in the seeds of plants, and in bones.

Pure phosphorus is a transparent solid, easily cut at ordinary temperature, and exhibiting a waxy appearance, unless exposed to the light, when it turns yellow, then red. It possesses the property of giving out light in the dark, and hence its name, which means *light-bearer.*

Phosphorus is a violent poison, and, except in very small quantities, a dangerous substance to use; for in the air it takes fire with the slightest friction, and burns with the greatest energy. It should be kept and cut under water, and, when taken out for use, should not be touched with the naked hand.

Silicon is not found free, but always combined with oxygen, forming silica or quartz, a mineral which will be noticed in a subsequent chapter.

Aluminium occurs only in combinations. With oxygen it forms alumina, pure examples of which are the gems known under the name of "ruby" and "sapphire." Alumina enters largely into the composition of feldspar and clay.

Aluminium is now obtained in considerable quantities from cryolite, a mineral from Greenland, and from clay. It is a white metal, resembling silver, is very light, and is not tarnished by exposure.

Potassium is one of the lightest of the metals. It occurs only in combinations. With oxygen it forms potassa, or potash, which is found in all fertile soils, and in all hard-wood trees, from the ashes of which it may be readily obtained.

Sodium forms more than 40 parts in 100 of common salt, and occurs in various other combinations in the earth's crust.

Calcium, always combined with oxygen in nature, occurs in every variety of limestone.

Magnesium combined with oxygen forms magnesia, which enters into the composition of many rocks and minerals, as will be seen in the next chapter.

Iron and Manganese will be considered in the next chapter.

CHAPTER IV.

MINERAL CONSTITUTION OF THE EARTH.

A SIMPLE mineral is either an element, or a union of two or more elements; and minerals associated together so as to form a homogeneous mass, constitute *rocks*. Thus, gold is an element, and at the same time a simple mineral, while two elements are chemically combined to form the simple mineral quartz, and three minerals are mixed together in the rock called granite.

There are about 600 simple minerals. To treat of these in all their varied forms is the special province of Mineralogy; but a general knowledge of minerals is necessary to the student of Geology; therefore this chapter is devoted to a description of those which are most abundant.

Although the number of minerals is so large, eight or nine constitute the great bulk of the earth. These are Quartz, Feldspar, Mica, Limestone, Hornblende, Serpentine, Gypsum, Talc, and Oxyd of Iron.

QUARTZ, Silica, or Silex, is the most abundant of all minerals, constituting nearly half the crust of the earth.

MINERAL CONSTITUTION OF THE EARTH. 27

It is one of the constituents of granite, and enters largely into the composition of other rocks, and occurs in abundance, in the form of sand, in nearly all soils.

Quartz is composed of silicon and oxygen—a metalloid and a gas. It is hard, scratches glass with facility, cannot be cut with a knife, breaks into irregular fragments under the hammer, and, with the exception of the fluohydric, is unaffected by acids.

It is proper to observe here that the following minerals constitute what is called the scale of hardness, the number appended to each name indicating the relative hardness of that mineral.—*Talc* 1, *Rock Salt* 2, *Calc Spar* 3, *Fluor Spar* 4, *Apatite* 5, *Feldspar* 6, *Quartz* 7, *Topaz* 8, *Sapphire* 9, *Diamond* 10.

With these minerals all others are compared. If a mineral is no harder than talc, its hardness is said to be 1; if as hard as calc spar, its hardness is said to be 3; if as hard as quartz, 7, and so on. If a mineral can be scratched with quartz it is softer than the latter; if it is scratched with quartz, and will itself scratch feldspar, it is evident that its hardness is between 6 and 7. And the same rule holds good in all cases. In studying a mineral, the hardness is among the first things to be determined; it is therefore desirable to possess the minerals enumerated above, in order that we may use them in making comparisons.

Quartz is of every shade of color, owing to other substances which are combined, or mixed with it.

Flint is compact quartz, often found in nodules in chalk.

Rock crystal includes all the crystallized colorless varieties. The ancients believed this to be congealed water, and accordingly applied to it the term *krustallos*; hence our term crystal, now applied to all minerals "bounded by plain surfaces symmetrically arranged."

Every substance that passes from a liquid to a solid state crystallizes, and each species in a manner peculiar to itself. In fact, crystallization and solidification are one and the same; but those perfect forms which delight the eye of the mineralogist, are produced only when the particles are free to arrange themselves according to the law which governs them. In the cooling of a mass, as a mass of melted iron, the crystals intermingle in every possible manner, and hence, under such circumstances, no perfect crystals are formed. Ice is massive, but every snow-flake is a cluster of crystals, perfect in form and finish.

Crystallography is the department of Mineralogy which treats of crystals; and a knowledge of this and of Chemistry enables the patient learner to identify every mineral substance.

The number of fundamental mineral forms, that is, the forms from which all others are derived, is thirteen. These are prisms, octohedrons, and dodecahedrons. The number of derived forms is very great.

Quartz crystallizes in six-sided prisms, terminated

by six-sided pyramids. Figures 1—4 illustrate the common forms of quartz crystals.

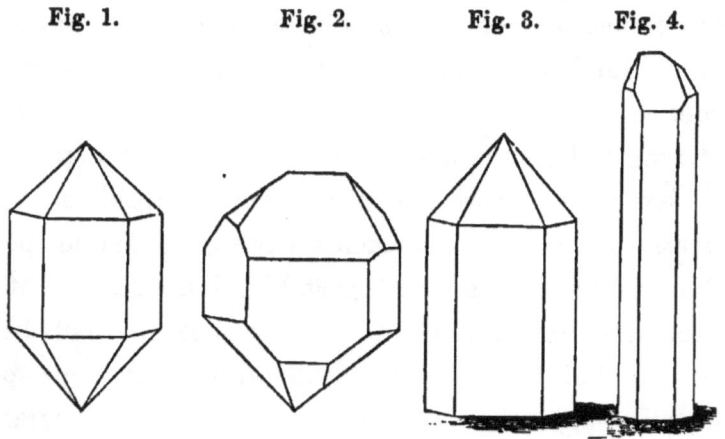

Some of the common forms of Quartz Crystals.

While these forms appear different at first glance, a closer inspection shows them to be alike in all essential particulars. Though the crystal be elongated, contracted, or flattened, or if one side be enlarged at the expense of another, the number of sides and the value of the angles are constant.

When, as at Little Falls, N. Y., and at many other localities, crystals occur loose in cavities, each end is generally terminated with a pyramid. Quartz crystals, however, are more frequently attached, as seen in the beautiful group from Warwick, Mass., represented by the accompanying wood-cut.

Fig. 5.

Group of Quartz Crystals, Warwick, Mass.

Geodes, or hollow pebbles, with the concave surface thickly studded with quartz crystals, are abundant in the limestone of many regions, and especially in the limestone bluffs along the Mississippi river. Externally these geodes are mere pebbles, but a blow from the hammer reveals great beauty within.

Pure crystals of quartz, and of every other mineral, are generally small. Quartz crystals of inferior clearness are sometimes found of enormous dimensions. A group in the Cabinet of Dartmouth College weighs 147 pounds, and some of the crystals are five inches in diameter. There is a group in the University of Naples which weighs half a ton. At Milan there is a crystal of quartz three feet and three inches in length, six feet in circumference, and weighing 800 pounds.

Rock crystals, in greater or less perfection, occur

almost everywhere. Canada Creek, St. Lawrence county, N. Y., Little Falls, in the same state, and Smithfield, R. I., are among the noted localities in the United States. The largest crystals in the world come from the Alps.

The purest specimens of rock-crystal are often cut and used in jewellery, optical instruments, and spectacle-glasses. Cups and vases were formerly cut from this substance. Quartz crystals containing rutile afford beautiful gems.

Amethyst is a purple variety of crystallized quartz. It is colored by the oxyd of manganese, or by iron and soda. It was named by the ancients, who believed that wine drunk from goblets made of this mineral would not intoxicate; and this idea is expressed in its name.

Amethyst has always been esteemed as a gem, but is more brilliant by sunlight than by gaslight. Fine cabinet specimens are found in Lincoln county, North Carolina. Small crystals are plentiful on Mount Holyoke. The finest varieties used for jewellery are brought from Brazil, Ceylon, India, Siberia, and from various parts of Europe.

Rose quartz is a massive variety, of a rose or pink color, which is probably due to manganese. Beautiful specimens can be obtained in abundance at Acworth, N. H., and at Paris, Me.

Prase is a leek-green, massive variety of quartz.

Smoky quartz occurs both crystallized and massive. Large crystals have been found in the vicinity of the White Mountains, N. H.

Ferruginous quartz is yellow, or brown red, the color being due to the oxyd of iron.

Chalcedony is a gray or white variety of quartz, with a lustre of wax, and often exhibiting a botryoidal surface. It occurs in cavities of the rocks, where it was probably formed by the infiltration of silicious waters.

Carnelian is the same material as the last, with a bright red color. It is much used for the more common jewellery.

Chrysoprase is a leek-green variety of chalcedony.

Sard is a deep brown red variety of chalcedony.

Agate is a variety of chalcedony, in which the colors are arranged in clouds, spots, bands, and concentric lines. When the lines are zigzag, it is called *Fortification agate*. Sometimes moss-like delineations are found in agate, then it is called *Mocha-stone*, or *Moss-agate*. When the colors of the agate are arranged in horizontal layers, or bands, it is called *Onyx*, and is the stone used in making real cameos. When the layers are sard and white chalcedony, it is called *Sardonyx*.

Chalcedony, carnelian, and agate pebbles are abundant on the shores of Lake Superior, and other lakes of the West, also on the shores of Scotland.

Jasper is silica containing oxyd of iron and clay. It occurs of various colors, though it is generally red. Jasper is exceedingly abundant in Saugus, Mass.

Bloodstone, or Heliotrope, is a deep green variety of quartz, with blood-red spots. It contains some clay; and the spots are probably due to iron. It is much used for ring-stones and seals. Fine specimens come from India and Siberia.

Granular quartz is of a granular texture, resembling loaf-sugar.

Quartz is an important mineral in the arts. Window-glass is made of quartz, soda, and lime. Plate-glass of quartz, potash, and lime. Flint-glass and crystal-glass are made of quartz, potash, and soda. Green bottle-glass of quartz, alumina, oxyds of iron and manganese, and potash, or soda.

The great abundance of quartz is significant of its importance in the economy of nature. It is essential to the growth of vegetation. Many kinds of plants cannot exist without it. This is especially true of grasses and grain. It is this which gives firmness to their stems. When there is a deficiency of available silica, the grass or grain falls down. Some rushes contain so much silica that they are used for polishing purposes.

FELDSPAR enters largely into the composition of the earth's crust. It differs from quartz in having a regular cleavage, a pearly lustre, and in being somewhat softer than the latter. Its hardness is 6. Feldspar is composed of silica 64.20, alumina 18.40, potash 16.95. It is generally of a gray or whitish color; though flesh-red, green, and blue varieties are not uncommon in some regions. It is one of the constituents of granite. Feldspar is quarried at Acworth, Alstead, and Grafton, N. H.

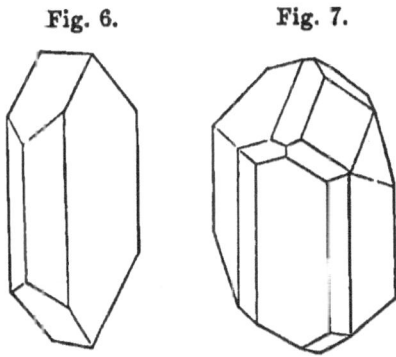

Fig. 6. Fig. 7.

Common forms of Crystallised Feldspar.

Feldspar is extensively used in the manufacture of mineral teeth and porcelain. When decomposed, it forms a clay called *Kaolin*, used in the manufacture of pottery. Common brick-clay is impure decomposed feldspar, and the bricks owe their red color to the presence of iron in the clay.

A variety of feldspar, containing a large proportion of soda, is called *Albite*, and is easily distinguished from the common variety by its invariable whiteness.

Moonstone and *Sunstone* are varieties of feldspar sometimes used as gems.

MICA, with quartz and feldspar, constitutes granite. When the constituents are coarse, we have coarse granite, but when fine, they form the compact granite suitable for building purposes. In the coarse granite rocks of Alstead and Grafton, N. H., plates of mica are sometimes obtained from one to two feet long, and a foot wide.

The elasticity of mica, and the remarkable facility with which it may be split into extremely thin leaves, readily distinguish it from all other minerals. Its colors are white, green, brown, yellow, and black. Its hardness varies from 2 to 2.5. Sometimes, as at Goshen, Mass., it is found of a beautiful rose or pink color, and hence is called *Rose Mica*. A beautiful variety found at Paris, Me., consisting of an aggregation of crystals of a pink or purple color, is called *Lepidolite*. A variety, consisting of black, globular, or oval concretions, from a half inch to two inches in diameter, is found at Newfane, and at Craftsbury, Vt.

Fig. 8.   Fig. 9.

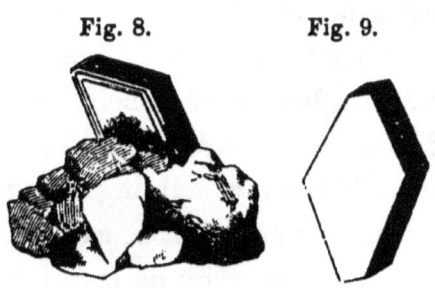

Common forms of Mica Crystals.

The plates of mica, which we observe more frequently than crystals, are but parts of crystals more or less regular. Mica has many important uses. Not being easily injured by fire, it is used in the doors of stoves, and in all places where a transparent or translucent substance is necessary, and at the same time where one is required that will not easily break. The composition of common mica is silica 46.3, alumina 36.8, potash 9.2, oxyd of iron 4.5, fluoric acid 0.7, water 1.8.

Fig. 8 is a perfect copy of a beautiful crystal with its lower portion embedded in rock, from Gilsum, N. H.

LIMESTONE, or Carbonate of Lime, although a simple mineral, occurs in vast masses, forming the flanks of mountains, and underlying extensive regions. It is readily distinguished from quartz, by its inferior hardness—being easily cut with a knife—and by its effervescence with acids. Its hardness is 3. Its composition is lime 56.3, and carbonic acid 43.7. Dolomite is a carbonate of lime, containing magnesia.

Considered in regard to structure, limestones are either *compact* or *granular*. The compact limestones break with a smooth surface, often conchoidal. The granular are crystalline, some varieties resembling loaf sugar.

Limestones capable of taking a polish, are called

Marble, which is familiar to all in the form of table-tops, floors, mantel-pieces, fire-jambs, &c.

The purest and best crystalline limestone is used in sculpture, and is called statuary marble, the finest qualities of which come from Carrara in Italy, and from the island of Paros. The celebrated Parian marble, used by the Grecian sculptors, came from the latter place.

Marble has always been extensively used for architectural purposes. Many of the ancient temples were built of it. The Parthenon at Athens was built of white marble. It occurs, suitable for building purposes, abundantly in Vermont, Western Massachusetts, Eastern New York, and Western Connecticut.

Many varieties of marble are beautifully shaded and mottled. These markings are due to foreign matter, and, in many instances, to the shells, corals, and encrinites which are scattered through the rock, and not unfrequently constitute its main bulk.

Chalk is a carbonate of lime. *Calcareous Tufa* is vegetable matter incrusted or petrified by deposits of carbonate of lime from lime springs. Such springs are common. One in Williamstown, Vt., has formed a deposit of calcareous tufa of considerable extent.

In caves, calcareous waters form *Stalactites*, which hang from the roof like icicles; and the water which drips from a stalactite deposits a mass upon the floor more or less conical, called *Stalagmite*. Not unfrequently beautiful columns are formed in this way.

Fig. 10.

Cave with Stalactites, Stalagmites, and Columns.

The crystallized varieties of limestone are all known under the general name of *Calc Spar*. More than seven hundred crystallized forms of this mineral—all modifications of the rhombohedron—have been described by mineralogists. The primary form, the rhombohedron, and some of the most common modifications, are represented by the figures on the following page. Fig. 14 is the form of the variety known as Dog-tooth Spar, very abundant at Lockport, N. Y.

All varieties of calc spar cleave into rhombohedrons, transparent specimens of which have the property of double refraction—of making objects over which they are placed appear double. Such specimens are familiarly known under the name of Iceland Spar.

MINERAL CONSTITUTION OF THE EARTH. 39

Fig. 11.   Fig. 12.   Fig. 13.   Fig. 14.

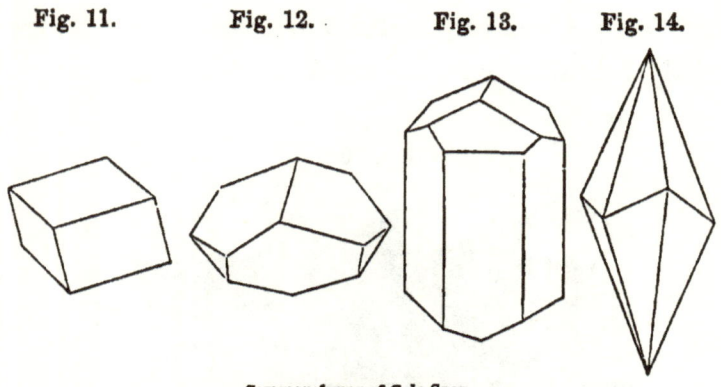

Common forms of Calc Spar.

The crystallized varieties may be found in all limestone regions. The finest specimens in the United States occur in St. Lawrence county, N. Y. Beautiful specimens may be obtained at every lime-quarry.

When burnt, every variety of limestone yields quick-lime. Heat drives off the carbonic acid, and leaves the lime.

Besides the various uses of limestone for building purposes, it performs important offices in the vegetable and in the animal kingdom. It is indispensable to the pod-bearing plants, and many others. It constitutes the chief material of the hard parts of many of the lower animals.

Coral reefs, composed of the skeletons of polyps— to be explained in a subsequent chapter—are mostly carbonate of lime. So also are the shells of the oyster, clam, lobster, crabs, and all other so called shell-fish.

MINERAL CONSTITUTION OF THE EARTH.

HORNBLENDE is a tough mineral, generally dark-colored, though it is found of all shades, from white to black. It occurs everywhere in the rocks of volcanic origin, in some of the older slates, and in syenite, which will be noticed hereafter. Its hardness varies from 5 to 6. Tremolite, actynolite, and asbestus, are varieties of hornblende. *Asbestus* occurs both compact and fibrous. It is often erroneously pronounced to be petrified wood. Sometimes it is so fibrous that it can be spun and woven like cotton. The ancients made napkins of it, and, as it is incombustible, they cleansed them by throwing them upon the fire. It was also used for wicks in the lamps of the temples; and because it was not consumed, the mineral was called *asbestus*, which name expresses that fact. Asbestus is abundant in Richmond county, New York, and in many other places in the United States.

Composition of hornblende: silica 48.8, magnesia

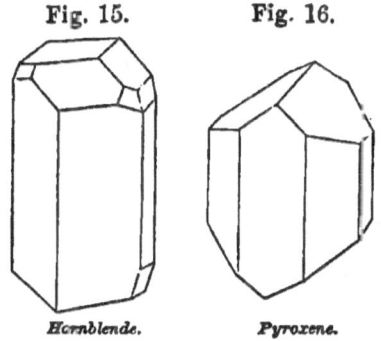

Fig. 15. Fig. 16.
Hornblende. Pyroxene.

Common form of Hornblende and of Pyroxene Crystals.

13. lime 10.2, alumina 7.5, protoxyd of iron 18.75, protoxyd of manganese, hydrofluoric acid, and water, 1.24.

Pyroxene, under the various names of Diopside, Sahlite, Augite, &c., closely resembles hornblende, but differs from it in cleavage, and in the angles of its crystals.

TALC sometimes resembles mica, from which it may be readily distinguished by its inferior hardness, and by its inelasticity. Talc is one of the softest minerals, has a foliated structure, and feels very unctuous. Massive talc is called *Steatite*, or Soapstone, and is in common use for stoves, stove and fire-place linings, &c. It is extensively quarried in New England. Groton, Middlefield, Windsor, Blanford, Chester, and Andover, are the principal localities in Massachusetts; Orford and Francestown in New Hampshire; and Grafton and Windham in Vermont. Talc and its varieties are mostly silica and magnesia.

SERPENTINE occurs massive, often forming extensive hills, as at Cavendish, Vt. Its colors are shades of green. Frequently it is found beautifully clouded, presenting a fine appearance when polished. It is composed of silica 42.3, magnesia 44.2, protoxyd of iron 0.2, carbonic acid 0.9, and water 12.4. Its hardness varies from 2.5 to 4.

Precious Serpentine is of a rich oil green color, and is a good material for inlaid work. This variety occurs near Newburyport, Mass.

The clouded varieties are well known under the name of *Verd-antique*, extensively used for tables, fire-jambs, and various ornamental purposes. This variety is quarried of great beauty at Roxbury, Vt.

GYPSUM, or Sulphate of Lime, constitutes extensive formations in New York, the Western and South-Western States, and in Nova Scotia. It is also abundant in Europe. Its composition is lime 32.9, sulphuric acid 46.3, and water 20.8. It is readily distinguished from carbonate of lime by being softer, and by not effervescing with acids. Its hardness is about 2. When ground, it is used as a fertilizer. It is also extensively employed in taking casts and models, and for hard finish on the walls of houses. For these purposes the gypsum is ground and heated, to expel the water; then, by wetting it again, it is made into paste, which hardens almost immediately. Gypsum is often found in beautiful crystals. Sometimes the crystals are double, as in Fig. 18.

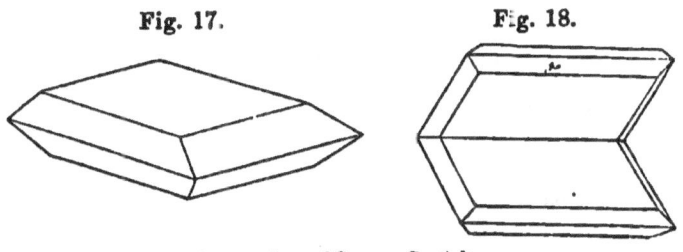

Fig. 17.   Fig. 18.

Common form of Gypsum Crystals.

When found in foliated masses, it is called *Selenite*,

and can be readily split into plates like mica, but not to the same extent. It is much softer than mica, and inelastic. Compact gypsum of a fine grain, and of a pure white, or light color, is called *Alabaster*, and is much used for vases and various ornaments. Castalino, Italy, furnishes the finest alabaster. In the Mammoth Cave, Kentucky, it occurs in beautiful imitations of vines, flowers, and fruits.

OXYD OF IRON is widely diffused, occurring in nearly all soils, and disseminated among the solid rocks, and often covering them in the form of rust. It also occurs in veins and beds, and sometimes constitutes the great bulk of mountains. More extended remarks upon iron are made under the head of Metals and Metallic Ores.

OTHER MINERALS OF IMPORTANCE TO THE GEOLOGIST.

Although the minerals noticed in the previous pages are the most abundant, it is important that the student become acquainted with others which he will frequently meet with in his investigations, and which are worthy of attention, both on account of their intrinsic value and interest, and the aid they will afford in the examination of the rocky formations of our globe.

The additional minerals which will be briefly noticed, are Heavy Spar, Celestine, Fluor Spar, Apatite, Scapolite, Andalusite, Staurotide, Kyanite, Garnet, Idocrase, Epidote, Chondrodite, Spinel, Corundum, the Zeolite Family, Tourmaline, Iolite, Beryl, and Zircon, and the most common Metals and Metallic Ores.

HEAVY SPAR, or Sulphate of Baryta, is a white or gray mineral, often associated with metallic ores. It is readily distinguished from the minerals which it resembles, by its greater weight. It occurs in laminated masses and in crystals. Its hardness varies from 2.5 to 3.5.

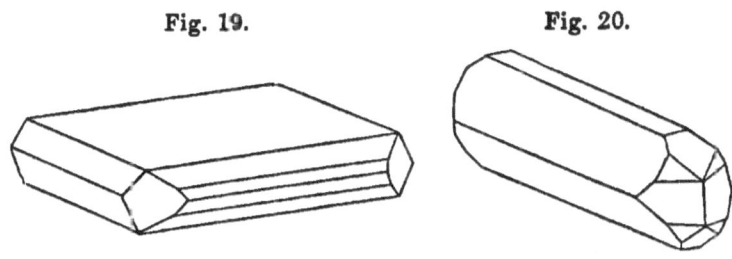

Common forms of crystallized Heavy Spar.

Heavy Spar is found in abundance at Hatfield, Mass., and Cheshire, Conn.; also in the Southern and Western States. Its composition is sulphuric acid 34, and baryta 66. It is used as a pigment with white lead.

CELESTINE, or the Sulphate of Strontia, occurs in long, slender, and often flat crystals, generally of a blue color, and readily distinguished from heavy spar by being lighter than that mineral; and from all the carbonates by not effervescing with acids. Its hardness is about 3.5. It is found on Strontian Island, Lake Erie; at Lockport and Rossie, N. Y. At Lockport it is associated with dog-tooth spar. Composition, sulphuric acid 43.6, strontia 56.4. Celestine is used for making the nitrate of strontia, the substance which gives the red color in fireworks.

FLUOR SPAR, or Fluate of Lime, is found both massive and crystallized in cubes, which readily cleave so as to form octahedrons. Its colors are dingy white, purple, green, brown, yellow, rose, and slate; but purple and green varieties are the most common. The purple resembles amethyst, but its inferior hardness easily distinguishes it from that mineral. Its composition is fluorine 47.7, calcium 52.3.

Fluor Spar occurs in large crystals at Muscalonge Lake, St. Lawrence county, N. Y. It is very abundant in Derbyshire, England, and hence is often called Derbyshire Spar. It takes a good polish, and is much used in England in the manufacture of vases, and various kinds of ornamental work. Fluoric acid, used in etching, is obtained from this mineral.

APATITE, or Phosphate of Lime, occurs both massive and in six-sided prisms. It is found in granular limestone, gneiss, and mica slate. Its softness will readily distinguish it from beryl, which it often resembles. It is easily scratched with a knife, dissolves in acids without effervescence, and phosphoresces when heated. Its colors are various, but the crystallized varieties are generally some shade of green. Its composition is phosphate of lime 92.1, fluorid of calcium 7.0, chlorid of calcium 0.9.

This mineral is an important fertilizer, entering into the composition of the kernel of the most important grains. Bones contain more than 50 per cent. of phosphate of lime. These facts show the important relation this mineral sustains to the vegetable and the animal kingdom. Beautiful crystals of apatite are obtained in St. Lawrence county, N. Y.

SCAPOLITE occurs in the form of square prisms, in granular limestone and in quartz. It also occurs massive. Its hardness varies from 5 to 6. Scapolite occurs in abundance at Governeur, N. Y., and at Bolton, Mass. At the latter place a massive variety is found of a beautiful lilac color, and hence is called Lilac Scapolite.

ANDALUSITE, also called Chiastolite, and sometimes Macle, usually occurs in clay slate, in four-sided prisms, more or less modified. The hardness of this mineral varies from 3 to 7.5. Cross sections of crystals

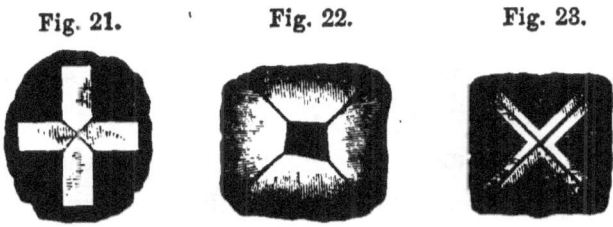

Fig. 21.   Fig. 22.   Fig. 23.

Perfect copy of cross sections of Andalusite Crystals, Lancaster, Mass.

exhibit something that reminds one of mosaic work, as may be seen in the accompanying figures, copied from nature.

STAUROTIDE occurs abundantly in the mica slate rocks of New England. Perfect specimens may be obtained in almost any quantity near Mink Pond, Lisbon, N H. Staurotide is also abundant in Charlestown in the same state. Its hardness varies from 7 to 7.5.

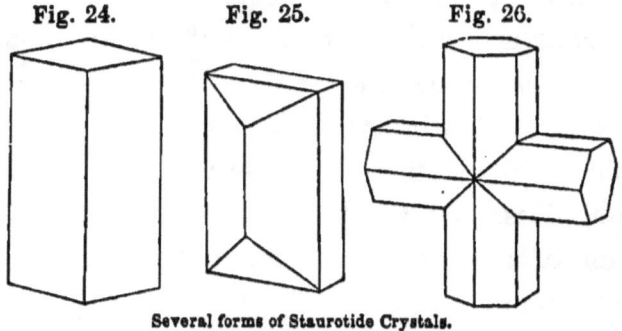

Several forms of Staurotide Crystals.

Crystals of this mineral are sometimes found, which exhibit a tesselated appearance like andalusite. According to Dr. C. T. Jackson, such are found where mica slate, containing them, passes into clay slate.

KYANITE occurs in long flat crystals, either single or aggregated, penetrating quartz, gneiss, and mica slate. Its color is sky blue, and when fresh from the quarry, the specimens are remarkably beautiful. Hardness 5 to 7. Kyanite occurs at Litchfield and Washington, Conn.; Chesterfield, Mass.; and Acworth, N. H.

GARNET occurs in dodecahedrons and double dodecahedrons. It is found of almost every color. Hardness from 6.5 to 7.5. *Precious Garnet* is deep red, and is transparent or translucent. Its composition is silica 42.5, alumina 19.15, protoxyd of iron 33.6, protoxyd of manganese 5.5. Common garnet is dingy red. *Cinnamon Garnet*, or Cinnamon Stone, is cinnamon brown, and hence its name. The composition of cinnamon garnet differs from the precious garnet, in containing only 5 or 6 per cent. of iron, and by having 30 per cent. of lime, while the precious garnet has no lime. Manganesian garnet is of a deep red color, and very brittle. It contains 30 per cent. of the protoxyd of manganese.

Fig. 27.   Fig. 28.   Fig. 29.

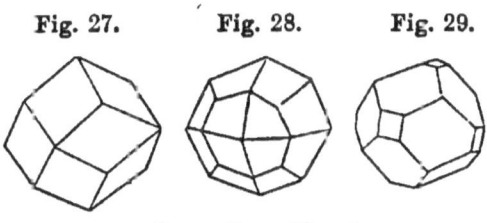

Common forms of Garnet.

Garnet occurs abundantly in mica slate, gneiss, and in granite, especially in the coarse varieties of the latter. Cinnamon garnet is generally found at the junction of limestone with silicious rocks. This mineral is found under such circumstances at Amherst, N. H., and at Carlisle, Mass.

IDOCRASE, or Egeran, resembles some varieties of

cinnamon garnet, with which it is often found; but differs in crystallizing in square prisms. Its color is brown, and its composition silica 37.4, alumina 23.5, protoxyd of iron 4.0, lime 29.7, magnesia and protoxyd of manganese 5.2. Hardness 6.5. Fine spe-

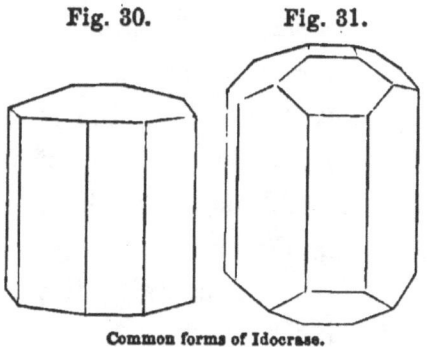

Common forms of Idocrase.

cimens are found at Parsonsfield, Me., and at Amherst, N. H.

EPIDOTE may be distinguished by its yellowish green color. It often occurs in beautiful six-sided prisms, but frequently it appears as a green coating

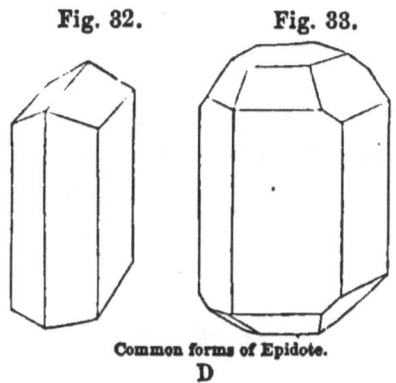

Common forms of Epidote.

upon the sides of fissures in the rocks, as at Nahant, Mass. It is generally found in the greatest perfection where dikes cut across rocks abounding in lime. It is found in fine crystals at Franconia, N. H. Its composition is silica 37.0, alumina 26.6, lime 20.0, protoxyd of iron 13.0, protoxyd of manganese 0.6, water, 1.8. Hardness from 6 to 7.

CHONDRODITE occurs in small brownish rounded kernels in granular limestone. Its composition is silica, magnesia, protoxyd of iron, and fluorine. Hardness 6.

SPINEL occurs in octahedrons and dodecahedrons, in limestone and gneiss. Its colors vary from red to black; blue and brown being common. Its composition is mainly alumina and magnesia. Hardness 8. This mineral and the last are abundant in the limestone from Amity, N. Y., to Andover, N. J. Small crystals of spinel are quite numerous in the limestone at Bolton, Mass. Clear crystals of this mineral are much used as gems. The red variety is the common ruby of the shops.

CORUNDUM occurs in hexagonal prisms, and massive. It is readily distinguished by its hardness, scratching all other minerals except the diamond. Blue is the prevailing color, but red, green, and yellow are not uncommon. The pure varieties, when blue, are called *Sapphire*, and when red, *Oriental Ruby*, and are highly prized as gems. The granular varieties, which generally contain some iron, are called *Emery*. The emery of commerce comes mostly from Turkey and the Grecian Islands. Corundum is found in many places in the United States, especially in New Jersey, Pennsylvania, North Carolina, and Georgia. The finest sapphires and oriental rubies come from India, especially from Ceylon.

The Zeolite Family comprises a great number of minerals, which are found mostly lining cavities and seams in rocks of volcanic origin. They are called zeolites, from the Greek *zeo*, to boil, because, when heated before the blow-pipe, they swell or boil. They are mostly composed of silica, alumina, and lime, or soda. The best localities of this family in this country, are Bergen Hill, N. J., the Copper Region of Lake Superior, and Peter's Point and Cape Blomidon, Nova Scotia.

The principal zeolites common in this country, are *Heulandite* a white or reddish mineral, in rhomboidal prisms; *Stilbite*, generally white, in elongated rectangular prisms; *Apophyllite*, generally white or grayish, in square prisms, with a sharp pyramidal termination; *Laumonite*, generally massive, with a radiating structure, and of a white or grayish color; *Natrolite*, in slender rhombic prisms; *Thomsonite*, consisting of masses of radiating fibres, or acicular crystals of a snow-white color; *Analcime*, in trapezohedrons of a milky color; *Chabazite*, in white or reddish somewhat cubical crystals; and *Prehnite*, generally found in reniform and botryoidal masses of a light green color. This mineral takes a handsome polish, and is used for various ornamental purposes. To this list we may add *Datholite*, which is reckoned among the zeolites by many writers. It occurs in rhombic prisms, generally of a white color. It forms a jelly with nitric acid.

TOURMALINE is a mineral of very common occurrence in granite, gneiss, mica slate, quartz, chlorite slate, and steatite. It is also found in granular limestone. Black and brown varieties are the most common. It also occurs blue, bright, and pale red, or pink, green, cinnamon brown, yellow, gray, and white. It crystallizes in 3, 6, 9, or 12 sided prisms. Hardness about 7.5.

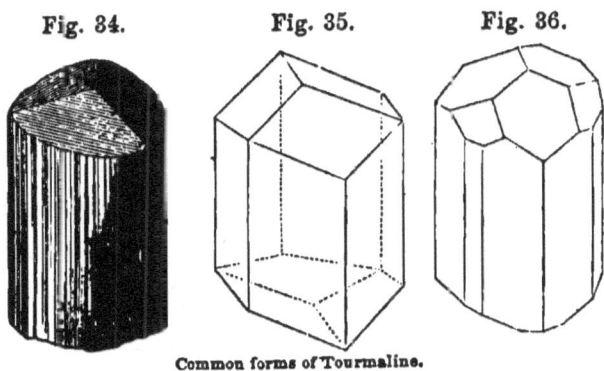

Fig. 34. Fig. 35. Fig. 36.
Common forms of Tourmaline.

The different colored tourmalines have received different names. Thus, Schorl was formerly applied to the black variety, but now it is called Black Tourmaline simply, while the term schorl is going out of use. Black tourmaline is often very highly polished, and beautifully striated. This is the case when it occurs in quartz, and the crystal leaves its impress upon the latter, even to the microscopic lines. The writer has many such crystals of great beauty from Sullivan, N. H.

The red varieties are called *Rubellite*, the blue and blue black, *Indicolite*. Black tourmaline has yielded silica 33.0, alumina 38.2, lime 0.8, protoxyd of iron, 23.8, soda 3.2, boracic acid 1.9. A specimen of rubellite yielded silica 39.4, alumina 44.0, potash 1.3, boracic acid 4.2, lithia 2.5, peroxyd of manganese 5.0.

Black tourmaline is found in all granite regions. The pink, red, and blue varieties occur in remarkable beauty at Paris, Me., and in less perfection at Goshen and Chesterfield, Mass. The red, pink, and green tourmalines of Paris, Me., when perfectly free from flaws, afford gems of great brilliancy and beauty.

IOLITE occurs mostly in hexagonal prisms, which are separable into layers at right angles to the length of the crystal. Its color is blue, of various shades. It occurs mostly in granite, gneiss, and talcose rocks. The principal localities in the United States are at Haddam, Conn., Brimfield, Mass., and Richmond, N. H. Pure iolite is cut for gems.

TOPAZ, one of the minerals mentioned in the scale of hardness, is found at Trumbull, Conn., the principal locality of this species in the United States. It is readily distinguished by its brilliant transverse cleavage. Topaz is generally pale yellow; but green, blue, and red shades are not uncommon. The finest specimens used in jewellery come from Siberia and Brazil.

BERYL occurs in six-sided prisms in coarse granite. Sometimes the crystal has flat terminations, as represented by Fig. 37, drawn from a specimen in the

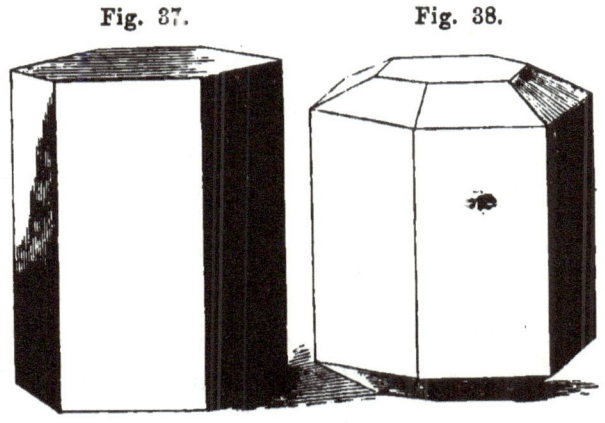

Common forms of Beryl.

author's collection, from Sullivan, N. H. More frequently no regular termination is found. Again, the crystal is terminated as in fig. 38, or in a modification of that form. The color of beryl is green, passing into blue, and sometimes into yellow. Clear beryl of sea-green color is called *aquamarine*, and is much used for gems. Beryl is composed of silica 66.5, alumina 16.8, glucina 15.5, peroxyd of iron 0.6. Hardness from 7 to 8.

Emerald is a rich green variety of beryl containing the oxyd of chromium, which gives it its color. When perfectly free from flaws, the emerald affords gems of great value. The finest specimens come from New Granada and Siberia.

The common beryl may be found in greater or less perfection in nearly all the coarse granite rocks of New England. Royalston, Mass., and Alstead, N. H., have furnished some of the purest specimens of any of our localities, but Acworth and Grafton, N. H., have furnished the largest. One crystal from Acworth is 4 feet long, and weighs 240 pounds. The writer has a large crystal of beryl found loose in the soil, on a hill in the south-west part of Marlow, N. H. The Royalston and Alstead beryls afford gems of great beauty.

The largest beryl in the world is embedded in quartz and feldspar on Mt. Alger, in Grafton, N. H. It is owned by Francis Alger, Esq., of Boston, to whom I am indebted for these facts about it. The entire length of the crystal is 9 feet; its circumference at the largest part nearly 12 feet; and its weight not less than 5 tons. In Mr. Alger's splendid collection, there is a crystal of beryl from this same locality, weighing $2\frac{1}{2}$ tons, some portions of which are transparent and very beautiful.

Chrysoberyl, found at Haddam, Conn., in tabular crystals, is closely related to beryl.

ZIRCON occurs in square prisms, terminated with four-sided pyramids. Its color is red, brown, or gray. Its hardness is 7.5. It is found in granite, gneiss, granular limestone, and volcanic rocks. Clear red specimens are called *Hyacinth*. Zircon is found at Litchfield, Me., Hammond, N. Y., Franklin, N. J., and at Green River, Henderson county, N. C. At the latter place the crystals occur loose in the soil, and in the greatest abundance. Clear specimens of this mineral are used for gems, and for jewelling watches.

METALS AND METALLIC ORES.

Metals are found either native or in the state of ores. A *native* metal is pure, or simply mixed with other substances, but not chemically combined with them. An *ore* is a metal chemically combined with one or more substances, such as oxygen, sulphur, carbonic acid, arsenic, silica, &c.

Metals occur in layers or beds; in veins intersecting the rocks; and disseminated through the rocks in grains and crystals.

The metals which will here be briefly noticed, are Iron, Manganese, Lead, Zinc, Copper, Rutile, Molybdenum, Tin, Silver, and Gold.

IRON is found native only in meteorites, those

wonderful bodies which occasionally fall to our earth. In combinations it is abundant, and widely diffused among the rock formations of our globe.

Oxyd of Iron, already noticed, is the most abundant and most important ore of this metal. There are three prominent varieties of oxyd of iron—the Magnetic, the Specular, and the Brown Iron Ore.

Magnetic Iron Ore occurs both in beds, and in octahedral and dodecahedral crystals. Vast beds of this ore occur in New Jersey, Pennsylvania, New York, New Hampshire, and in many other parts of the United States. The crystals are very common among the older rocks. At Amherst, N. H., the granite rocks abound with crystals of this mineral, which combine both the octahedron and the dodecahedron. This ore is extensively worked for the manufacture of iron. Its composition is iron 71.8, and oxygen 28.2. Hardness from 3.5 to 4.5. Lodestone is magnetic iron, which exhibits magnetic polarity. Such specimens are not unfrequently found in the iron regions.

Specular Iron Ore is so called from the high lustre of some of its varieties. Micaceous iron is a foliated variety of this ore. Some of the varieties do not exhibit lustre. Such are red hematite, red ochre, red chalk, and clay iron. Specular iron may be distinguished by its red color when pulverized. It occurs in great abundance in the rocks of all ages.

It abounds in Pennsylvania; and Pilot Knob and Iron Mountain, in Missouri, are mainly specular iron. At the island of Elba it is obtained in the most splendid crystals. The crystals have a hardness from 5.5 to 6.5.

This ore is extensively mined for the production of iron, but is less easily worked than the magnetic.

Brown Iron Ore is one of the most abundant and one of the most valuable iron ores in this country. It occurs both massive, and in nodular, botryoidal, and stalactitic forms; and is generally called Brown Hematite. Specimens often exhibit a beautiful iridescence. Brown and yellow ochre, and bog iron ore, are varieties of this mineral. Brown iron ore is abundant in Pennsylvania, New York, Massachusetts, and Connecticut; also in the Carolinas, and other states South and West.

There are many interesting iron ores besides the oxyds.

Spathic Iron, or Carbonate of Iron, is another important ore for the manufacture of both iron and steel. It often resembles calc spar, but may be distinguished from the latter by its greater weight. Its color is light brown. The composition of spathic iron is protoxyd of iron 61.37, and carbonic acid 38.63. Hardness from 3 to 4.5. Spathic iron abounds in the coal regions of Pennsylvania, and also occurs in New York, Vermont, and Connecticut. Specimens of this ore are readily obtained at Sterling, Mass.

Some idea of the vast amount of iron in the United States may be formed from the fact that the 1200 iron works in our country produce an aggregate of 850,000 tons annually.

Iron Pyrites, or Sulphuret of Iron, is disseminated in grains and crystals through the rocks of almost every region. It also occurs in extended masses, as at Strafford, Vt., where there is a vein four rods wide. It crystallizes in cubes and in modifications of the cube. This mineral somewhat resembles gold, for which it has been so often mistaken that it is familiarly known under the name *Fool's Gold*. It may be readily distinguished from gold by its greater hardness and brittleness, and by the sulphur odor which it gives off when heated. Hardness from 6 to 6.5. The composition of iron pyrites is iron 45.74, sulphur 54.26.

This species of iron ore is not worked for iron, on account of the difficulty of separating the iron from the sulphur; but it is extensively used for the manufacture of green vitriol, or copperas, sulphuric acid, and alum. Sulphur is also sometimes obtained from it.

Arsenical Iron Pyrites differs from the last in being of a silvery white color, crystallizing in prisms—though mostly found massive—and in giving off arsenical fumes when heated. It is composed of iron 36.0, arsenic 42.9, sulphur 21.1. Hardness from 5.5 to 6.

Chromate of Iron is an ore of a dark brown color, usually found in serpentine, and used in the manufacture of paint.

GALENA, or Sulphuret of Lead, occurs in granite, limestone, clay slates, and sandstones. It crystallizes in the form of cubes, and readily breaks into the same forms. The composition of galena is lead 86.55, sulphur 13.45. Hardness 2.5. This mineral generally contains more or less silver. The most extensive mines of this mineral in this country are at Galena, Ill., and the portions of Iowa and Wisconsin adjoining.

Phosphate of Lead is a variety crystallizing in six-sided prisms, or occurring in botryoidal masses. Its color is green, and the composition differs from galena in comprising phosphoric acid instead of sulphur.

Carbonate of Lead occurs both massive and crystallized. Its colors are brown and white. Both this mineral and the last are found in great perfection at the Wheatly mines, Phœnixville, Pa.

ZINC is never found in a native state, but combined with sulphur, oxygen, carbonic acid, or silica. *Sulphuret of Zinc*, or Blende, is of a wax yellow color, sometimes brown, and occurs with galena. By the miners it is called Black Jack. Hardness from 3.5 to 4.

Red Oxyd of Zinc occurs in foliated masses, of a bright red color. It is abundant at Franklin and Sterling, N. J., associated with calc spar. It is extensively mined for the manufacture of zinc and paint.

The silicate and carbonates of zinc are important ores for the production of the zinc of commerce.

COPPER occurs both native and as an ore. Native copper is found mostly near rocks of igneous origin. It occurs both in octahedral crystals and in irregular masses. Hardness from 2.5 to 3. Native copper occurs more abundantly in the vicinity of Lake Superior than in any other portion of the world. One mass discovered there was 50 feet long, 6 feet deep, and 6 inches thick.

Vitreous copper, gray copper, red copper, and malachite, are ores of this mineral.

COPPER PYRITES, or sulphuret of copper and iron, occurs both massive and in crystals. This ore resembles iron pyrites, with which it is generally associated, and, like the latter, it also resembles gold. It is, however, of a deeper yellow than iron pyrites, and much softer. The readiness with which it crumbles, and the sulphur fumes it gives off when heated, distinguish it from gold. Its composition is sulphur 34.9, copper 34.6, iron 30.5.

Sulphuret of copper is common in the United States, and is extensively mined in England and on the Continent. In Cornwall 12,000 tons of pure copper are obtained from this ore annually. Besides yielding copper, it is extensively employed for the manufacture of blue vitriol, or sulphate of copper.

Green Malachite, a green carbonate of copper, is a variety of great beauty and interest. It generally occurs in botryoidal masses—sometimes fibrous—in the copper regions, and is readily distinguished by its green color, and its perfect solution and effervescence with nitric acid.

Malachite originates from the decomposition of other ores of copper. It is slightly soluble in water, and has evidently been deposited as stalactites and stalagmites. Its composition is oxyd of copper 70.5, carbonic acid 18.0, water 11.

Malachite is found at Morgantown, Berks county, Pa., and in the copper-mines of the West. The finest specimens come from the Ural Mountains. A mass was found there 18 feet long and 9 feet wide, and estimated to weigh a half million of pounds. Malachite is much used for inlaid work and for jewellery.

RUTILE, or the Oxyd of Titanium, is a reddish brown mineral, with a metallic lustre, and usually in striated crystals, embedded in granite, gneiss, or mica slate. It is employed in coloring mineral teeth, and in painting upon porcelain.

Sometimes specimens of limpid quartz are found penetrated in every direction by long needle-like crystals of this mineral. Such stones, when polished, are very beautiful.

MOLYBDENUM occurs combined with sulphur, and deserves notice here, as it closely resembles some varieties of plumbago, but may be distinguished from the latter by its paler color, and by its sulphur fumes when heated. It occurs at Westmoreland, N. H., and in many other places.

TIN ORE, or Oxyd of Tin, occurs in veins in the

older rocks. It is found both massive and in crystals, which resemble those of black garnet. Except in Jackson, N. H., where there is reason to believe it occurs to a considerable extent, it is found only sparingly in the United States. Cornwall, England, Saxony, Austria, China, and the East Indies, comprise the great tin localities of the world.

SILVER occurs both native and in numerous combinations. Native silver is often found forming an alloy with gold; and is also frequently disseminated among native copper, especially that of the Lake Superior region. Native silver crystallizes in octahedrons.

Sulphuret of Silver, Sulphuret of Silver and Antimony, and Chlorid of Silver, are the principal ores worked for this metal in South America, Mexico, and Europe, whence most of the silver is obtained.

GOLD is always found native—with the single exception that it sometimes occurs combined with a rare substance, called Tellurium. Though forming no chemical combinations, it often occurs alloyed with silver, or copper, or both together.

Gold is found in grains, flakes, and masses, in the sands which the rivers bring down from the mountains; it is also disseminated in quartz veins.

Gold is readily distinguished from pyrites, with which it is often associated, by the facility with which it may be cut into slices, and flattened under the hammer. Its hardness is only 2.5 or 3. Pyrites, on the contrary, cannot be cut, and it crumbles under the hammer, and gives off sulphur fumes when heated. Gold crystallizes in octahedrons, and occasionally beautiful specimens are found.

Our limited space has allowed the statement of only some of the most important facts about the minerals noticed in the previous pages. For further information concerning them, and for a full description of others which are not mentioned in this treatise, the student is referred to Dana's works on Mineralogy, also to those of Shepard on the same subject.

In concluding this chapter, let me urge the importance of collecting specimens of all the minerals which can be found. For after all, it is only by collecting and studying the objects themselves that the student will make any considerable progress in becoming acquainted with the minerals of our globe. Although books may aid, the best descriptions fall far short of the reality. Let us go to the fields, to the mountains, to the quarry, to the canal and railway excavation, and there examine carefully every mineral substance, and secure a specimen of each for future study.

What a pleasure it is to bring to light those shining crystals which lie hidden in the earth upon which we tread! How full of instruction they are to him who studies them in the right spirit!

CHAPTER V.

THE ROCKS WHICH COMPOSE THE EARTH.

SECTION I.

TWO GREAT CLASSES RECOGNISED.

THE last chapter was devoted to a brief description of the most common minerals. This chapter describes the rocks of our globe.

The rocks which compose the earth may be divided into two classes.—the Stratified, and the Unstratified.

Fig. 39.

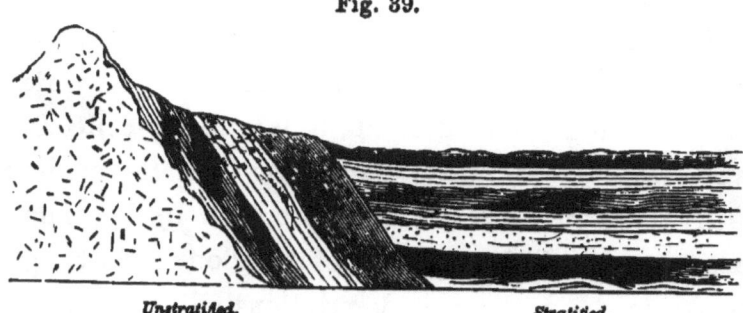

Stratified and Unstratified Rocks.

STRATIFIED ROCKS are those which occur in layers

or strata, parallel to one another, whether horizontal, inclined, or tortuous.

UNSTRATIFIED ROCKS do not exhibit layers, but are massive.

These two classes of rocks evidently had a different origin; or, at least, they have been subject to different influences. If we examine the loose materials, such as sand, clay, and gravel, that have accumulated at the bottom of a pool, pond, or lake, we find them arranged in beds or layers. The streams carry down the loose materials from the highlands, and deposit them in the water basins. Thus layer after layer is spread out, on the shores and bottoms of ponds, lakes, and oceans. Such deposits only need hardening to form stratified rocks. The ancient lakes and oceans were filled with layers of loose material, just as those of the present day are filling up. From these facts, and many others which will appear in subsequent pages, we may safely conclude that the stratified rocks have been formed in water basins, in the form of mud, sand, gravel, &c., and have been hardened by heat and other agencies. The stratified rocks are thus of *aqueous* origin.

On the other hand, examinations have shown conclusively that all the unstratified rocks have been melted, and therefore are said to be of *igneous* origin. Some of them are the result of the cooling of the original igneous mass; for it is believed that the whole earth was once in a melted state, and that its interior is now in that condition; others have been produced by the remelting of aqueous rocks; and others still have been poured out from the heated interior of the earth.

All the unstratified rocks which have resulted from the remelting of aqueous rocks, as well as those aqueous rocks which still retain their stratification, but have become crystalline, or otherwise much changed by heat, are called Metamorphic Rocks.

Undoubtedly the first solid rocks formed were unstratified, and resulted from the cooling of molten matter. But ever since water has occupied the surface of the earth—for the water must have been held in a state of vapor while the earth was in a molten condition—both stratified and unstratified rocks have been formed in every geological period. While water has been producing one class, heat has been producing the other. Hence, we not only find unstratified rocks beneath the stratified, but also cutting across both the stratified and the unstratified rocks, and sometimes in vast masses overlying the stratified rocks. The illustration at the close of the Seventh Chapter, shows the relative position of the two great classes of rocks mentioned above.

SECTION II.

DESCRIPTION OF THE UNSTRATIFIED ROCKS.

The principal unstratified rocks are Granite, Syenite, Porphyry, Greenstone, Basalt, Trachyte, Amygdaloid, and Modern Lavas. The varieties of each are very numerous.

GRANITE is composed of quartz, feldspar, and mica, promiscuously mixed together. It is coarse or fine, according to the state of its constituent minerals, which in some cases are so fine, that the different kinds are scarcely distinguishable; in others, the several constituents are in large irregular crystals or masses, a foot or more in diameter. Between these extremes, every possible grade may be found.

Granite is of every shade of color, but most commonly gray. When it contains distinct crystals of feldspar, no matter how irregular, it is called *Porphyritic Granite*, which is very abundant in many parts of New England.

Graphic Granite consists of quartz and feldspar, the former being generally dark-colored or smoky, and penetrating the feldspar in long irregular crystals or masses mostly parallel to one another. When the specimen is broken at right angles to the direction in which the quartz penetrates, it gives a surface somewhat resembling a page of written characters.

Fig. 40.

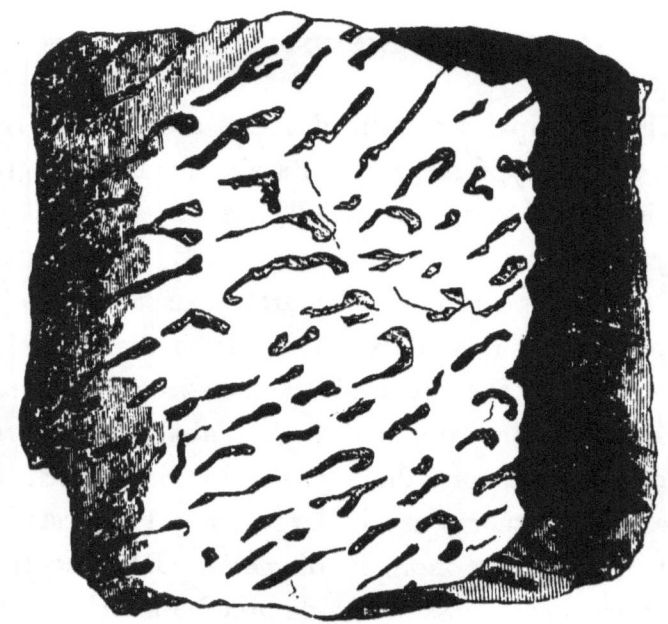

Perfect copy of a fragment of Graphic Granite, Alstead, N. H.

Granite is very abundant. It is one of the lowest rocks with which we are acquainted, and it also appears at the surface in mountain masses, and in veins traversing other rocks.

From the crystalline structure of granite, and the position in which it is found, there is no doubt but it has been melted. It is doubtful whether any of the granites which we are permitted to observe, are the direct result of the cooling of the original melted matter of our globe. On the other hand, it is quite certain that many of the granites have resulted from the remelting, and reconsolidation of stratified rocks. Therefore, while some granites may be the oldest rocks, others are more recent than some of the stratified rocks.

The relative age of granite, as well as that of all unstratified rocks, is to be determined by its position in respect to stratified deposits. Veins or dikes of granite, varying from an inch to many yards in width, are abundant, intersecting all kinds of rocks, and in all directions. Some of these undoubtedly are veins of segregation,—that is, accumulations of matter derived from the containing rock. Other veins have probably resulted from aqueous infiltration of the matter into an open fissure. Other veins still are fissures which have been filled with melted matter injected from the depths below, or furnished by some adjacent igneous mass. These last are called *veins of injection*, and are generally the best defined; but only extensive study in the field will enable the student to distinguish with certainty one kind from another.

Fig. 41 represents two granite veins in gneiss, Fairhaven, Mass. Here we have the record of at least three events. First, the formation of the gneiss; secondly, the injection of the small vein; and thirdly, the injection of the large vein right across the small one. It is plain that the vein which is cut off is older than the one which cuts it. Keeping these facts in mind, the relative ages of veins may be determined, no matter how numerous they are, or how much they intersect one another.

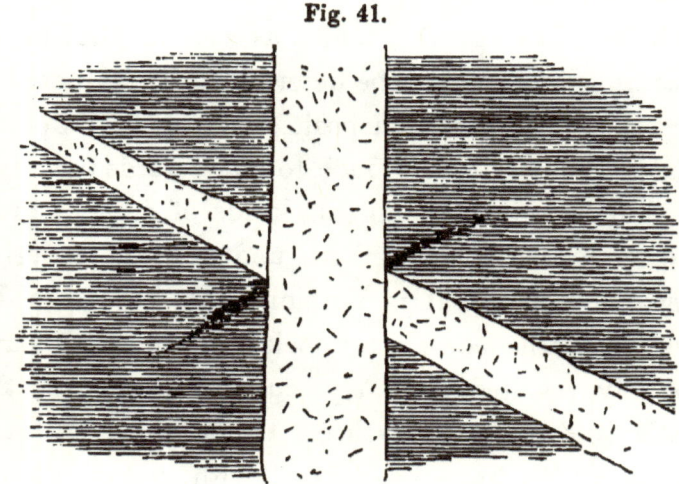

Fig. 41.

Two Granite Veins in Gneiss, Fairhaven, Mass.

The accompanying cut shows a vein or dike of injected granite, in hornblende slate, at Williams Hill, South Acworth, N. H. Near the road this vein is 15 or 20 feet wide, but towards the summit of the hill it widens, attaining a width of 90 or 100 feet. A little distance above, where the two persons are represented as standing, there is a large mass of the hornblende slate imbedded in the granite. This is indicated in the cut. The disturbance of the hornblende slate, which is considerably bent on both sides of the vein, and the position of the imbedded mass, leave little room to doubt that this hill was formed by the protrusion of the granite vein in a melted state. Similar cases are common in many parts of New England, especially in New Hampshire.

Fig. 42.

Granite Vein, Williams Hill, Acworth, N. H.

On the other side of this hill, just opposite where the horse and carriage are standing, is the celebrated beryl locality. The beryls occur in this vein of granite, but mostly beneath a vast mass of quartz.

Many interesting minerals are found in granite, as

the student has already noticed in our remarks on minerals in the previous chapter. Coarse granite almost everywhere yields beryl, tourmaline, and garnet in greater or less perfection.

Granite is extensively quarried for building purposes. It is very abundant, and of fine quality, in Massachusetts and New Hampshire.

SYENITE is composed of quartz, feldspar, and hornblende—the latter, however, being sometimes wanting. The term granite is often applied indiscriminately to this and the preceding rock. The celebrated Quincy Granite, so called, is syenite. Sometimes hornblende, as it occurs in syenite, resembles black mica, but may be distinguished from the latter by its brittleness, and by not separating into plates like mica.

The remarks made about the origin of granite are for the most part applicable to syenite. While some of the syenites may belong to the oldest rocks, others have resulted from the remelting of stratified deposits, and are thus true metamorphic rocks. Throughout the rocky portions of Eastern Massachusetts, the stratified rocks are found passing, by insensible gradations, into syenite, showing conclusively the metamorphic origin of the latter.

Syenite is very extensively employed for building purposes. The syenite of Quincy, Mass., may be found in the structures of almost every city in the

Union. The Custom House at Boston, and Bunker Hill Monument, are built of this rock.

PORPHYRY is a compact feldspathic rock, with crystals of feldspar disseminated through it. Porphyry occurs of various colors, but purple, brown, and green are the most common. The porphyry used by the ancients was purple. This rock takes a good polish, and is much used in some parts of the world for ornamental and architectural purposes. It is abundant on the coast of Massachusetts. The accompanying cut represents a fragment broken from a dike, near the Black Rock House, Cohasset, where good specimens are easily obtained.

Fig. 43.

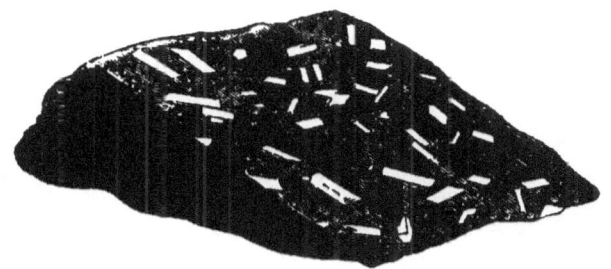

Fragment of Porphyry, Cohasset, Mass.

Porphyry pebbles of various color may be obtained in any quantity on the beach at Nahant.

GREENSTONE is a compact rock, composed mostly of hornblende and feldspar. Its color varies from dark green to brown or black. When hornblende predominates, it is sometimes called hornblende rock. It is found penetrating rocks of different ages, and in some cases it has flowed out over the surface. This, and the three following—Basalt, Trachyte, and Amygdaloid in all their varieties—are called *Trap Rocks*. Greenstone and basalt often assume a columnar structure, and they give rise to some of the most interesting scenery. Mt. Tom and Mt. Holyoke in Massachusetts, and the Palisades along the Hudson River, are greenstone.

Fig. 44 shows the columnar appearance on the side of Holyoke, where it slopes beneath the waters of the Connecticut. Professor Hitchcock long ago named this interesting place *Titan's Pier*.

Fig. 44.

Titan's Pier, Mt. Holyoke, Mass.

BASALT consists of augite and feldspar, and closely

resembles greenstone. It is compact, and of a dark color. It often exhibits a columnar structure. Fine examples occur along the Columbia river, in Oregon, and on the north shore of Lake Superior. Fingal's Cave, on the island of Staffa, and the Giant's Causeway, Ireland, afford wonderful exhibitions of columnar basalt.

Fig. 45.

Fingal's Cave, Island of Staffa.

In many cases the columns of basalt are jointed or divided into sections, usually with the upper part concave, and the lower convex, and fitting exactly together.

The columnar structure of the trap rocks is believed to be due to the tendency of the matter, when cooling, to assume the globular form. These globes press against one another, and the columnar structure is the result. The experiment of Watt, who melted 700 pounds of basalt, and allowed the same to cool slowly, seems to confirm this view of the case.

TRACHYTE is a gray, rough mineral, of volcanic origin. It is mainly composed of feldspar, with crystals of the same scattered through it. Sometimes it contains hornblende and mica.

AMYGDALOID is a term applied to any of the trap rocks, filled with cavities, either empty or containing quartz, spar, epidote, or other minerals.

Modern Lavas will be described in a subsequent chapter, when speaking of volcanoes.

All the trap rocks, and much of the porphyry, have been forced up, while in a melted state, from the heated interior of our globe. Veins or dikes of these rocks are abundant in many parts of the world. They occur from a fraction of an inch in width, to those which form mountain masses like Mt. Tom, and Holyoke, the Palisades, and the Island of Staffa, alluded to above. Trap and porphyry dikes are abundant on the coast of New England, especially in Massachusetts. They traverse the rocks in every direction at Cohasset, Nahant, Lynn, Salem, Beverly, Marblehead, and in many other places. One dike, near Pulpit Rock, at Nahant, is 34 feet wide. The following cut represents some very interesting examples in the metamorphic rocks at Cohasset.

Fig. 46.

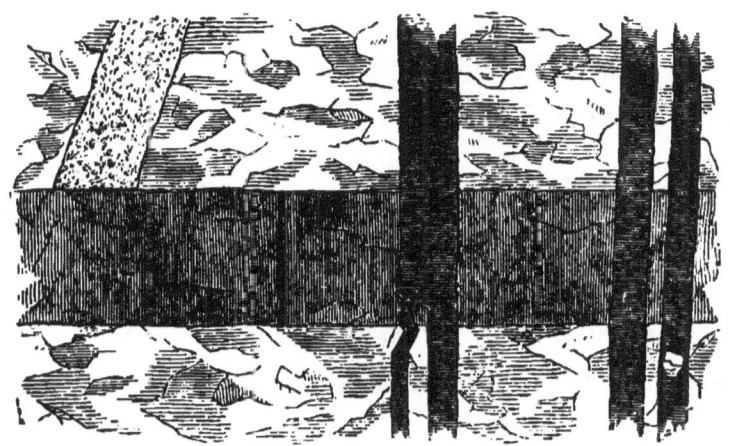
Dikes in Syenite, Cohasset, Mass.

The largest dike is 20 feet wide; the others vary from 5 feet to 10. The one on the extreme left is porphyry; the others are greenstone; the three smaller ones being very dark-colored on their weathered surfaces. A mass of the syenite is imbedded in the right-hand dike, as is indicated in the cut. Cases of this kind are not uncommon; and such a mass shows that it became imbedded while the dike was in a state of fusion. The mass itself generally gives evidence of having been fused upon its edges, thus, often passing into the matter of the dike by an insensible gradation.

In the case represented by Fig. 46, above, we have the record of several events. First, the formation of the syenite; secondly, the syenite was rent, and the fissure filled with the porphyry dike; thirdly, the syenite was rent across the porphyry, and melted rock flowed up forming a dike 20 feet wide; and fourthly, three fissures opened across the large dike, and melted rock flowed in and filled the rents.

These injections of melted matter change, more or less, the rocks through which they pass. It is probable that the metamorphic state of the rocks at Nahant, Cohasset, and many other places, is due in part, if not wholly, to the heating which they experienced during the formation of the dikes.

At the lime quarry in Rockland, Maine, a trap dike has changed the gray, coarse limestone near it, into white compact marble. Any number of similar facts might be cited; but, now that the attention of the student is turned to the subject, many cases will come under his own observation.

The attention has already been called to the relative ages of veins or dikes, and in closing this section, it is only necessary to explain one case, a little more complicated than those referred to in the previous pages, in order to prepare the student to begin successfully the study of this part of the subject in the field. Take the case illustrated by Fig. 47. The vein or dike *e* cuts off all the others, and is therefore the newest. The vein *d* cuts off *c*, and the latter cuts off *a* and *b*, and *b* cuts off *a;* therefore, *d* is the next older than *e*, *c* the next older than *d*, *b* the next older than *c*, and *a* is the oldest of the five veins. Thus, the relative ages of veins may be determined; but the length of time between the formation of the different ones cannot be found out.

Fig. 47.

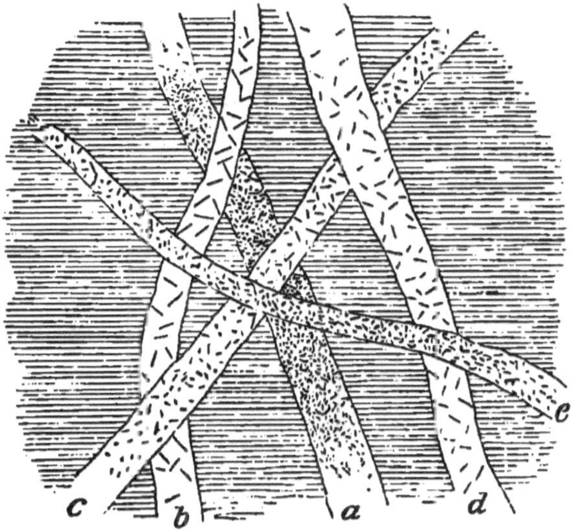

Relative Age of Veins or Dikes.

SECTION III.

DESCRIPTION OF THE STRATIFIED ROCKS.

After a brief description of the most common stratified rocks, some additional remarks will be made about the whole class. The most common stratified rocks are Gneiss, Mica Slate, Clay Slate, Hornblende Slate, Talcose Slate, Quartz Rock, Sandstones, Conglomerates, and Limestones.

GNEISS is often called stratified granite, because it appears like the latter, and is composed of the same minerals. Gneiss is very abundant in all parts of New England. It splits readily into slabs, and is much used for a building material, and for flagging-stones.

MICA SLATE resembles gneiss; the two being composed of the same minerals; and they often pass into each other by insensible gradations. Mica slate, however, has a more slaty appearance than gneiss, and generally has more mica, which often gives the rock a very glistening appearance when a fresh surface is exposed. It appears in numerous varieties, coarse, fine, loose, and compact. It is readily distinguished by its slaty structure, and by the scales of mica which abound in it.

Tourmaline, garnet, and staurotide, are common in this rock. At Warwick, Mass., a mica slate is filled with crystals of tourmaline, of great beauty. The crystals occur in stellate clusters. At Lisbon, N. H., the mica slate abounds with perfect crystals of garnet, and staurotide.

The firm varieties of mica slate are used for flagging-stones; and some of the fine varieties, which contain considerable sand, for scythe-stones.

HORNBLENDE SLATE differs from mica slate in containing hornblende instead of mica. In general appearance, it often resembles mica slate.

TALCOSE SLATE, in external appearance, resembles mica slate, but contains talc, instead of mica, which makes it feel soapy. When chlorite takes the place of talc, the rock is called chlorite slate, which is of a dark green color. Talcose rock is quartz containing talc, and constitutes the gold-bearing rock in many parts of the world.

CLAY SLATE, or argillaceous slate, is a stratified rock in which clay predominates, and which often composes almost the entire rock. The color is like common clay, or darker. *Roofing slate* is a variety, and is extensively quarried in Maine, Vermont, Massachusetts, and in various parts of Great Britain.

QUARTZ ROCK includes those stratified deposits which consist mainly of quartz, but which often contain more or less mica. The prevailing colors are gray, brown, and blue. Quartz rock and sandstones pass into each other by insensible gradations.

SANDSTONE is composed of grains of sand more or less firmly united. The great bulk of sandstone is quartz. The colors are various. A dark red sandstone, often called *Freestone*, is extensively used for building purposes. When firm sandstone splits readily, it is used for flagging.

CONGLOMERATE, or Puddingstone, consists of water-worn pebbles cemented together, thus forming a compact rock. Many kinds of pebbles are often associated in the same mass. In some cases they are small; in others, a yard in diameter, as at Newport, R. I.

LIMESTONES constitute extensive formations in all countries. All the varieties may be distinguished by their effervescence with acids. The important uses of limestone have already been pointed out.

As remarked on a previous page, the stratified rocks occur in layers or beds, more or less distinctly

Fig. 48.

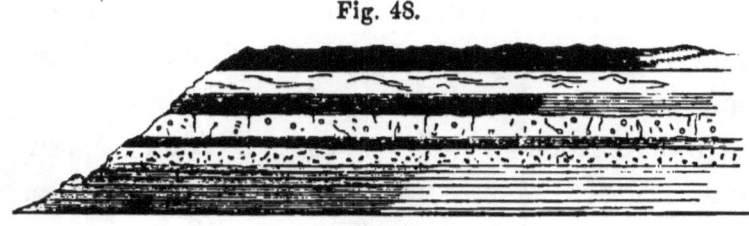

Horizontal Strata.

marked. These layers are called strata, and occur horizontal, inclined at all angles, and vertical; also, bent, folded, and contorted in every possible manner.

The exceedingly thin layers which stratified rocks often exhibit, are called laminæ. Sometimes they are easily separated; in other cases, they adhere with great firmness.

It is common to find both the stratified and the unstratified rocks traversed by divisional planes called *joints*. In stratified rocks, these planes are either vertical, or oblique to the planes of stratification. There are generally two sets crossing each other at angles more or less oblique, thus dividing the strata into blocks of considerable regularity. Examples of this structure are so numerous in all parts of our country, that it will not be necessary to illustrate it by a cut. These joints are of the highest importance to the quarryman, for they enable him to work a ledge to almost any extent desirable, without blasting; and the blocks, when taken from the quarry, are sure to have one or more smooth faces.

Joints traverse conglomerates just as well as other rocks, cutting right through the hardest quartz pebbles, as smoothly as through the softer portions.

When strata are inclined, the angle which they make with the plane of the horizon is called the *dip* of the strata. Thus the strata represented by Fig. 49 dip towards the north at an angle of 45 degrees.

STRATIFIED ROCKS. 85

Fig. 49.

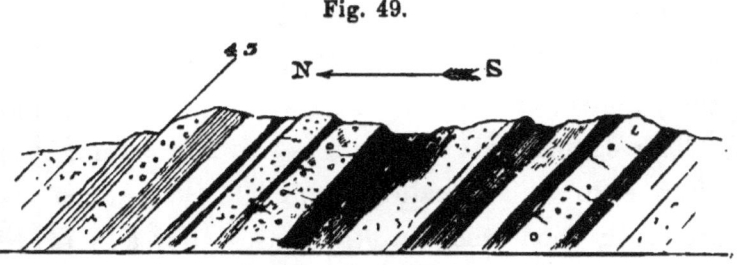

Strata dipping towards the North at an angle of 45°.

The direction of the upturned edges of strata, on a line at right angles to the direction towards which the strata dip, is called the *strike*. That is, if strata dip towards the north or the south, the strike is east and west. When perfect accuracy is required, geologists determine the dip by an instrument called the clinometer, and the strike by means of a compass. The dip may be determined near enough for all ordinary purposes by standing opposite the cliff, and placing the hands before the eyes in the position represented by Fig. 50, and observing whether the planes of the

Fig. 50.

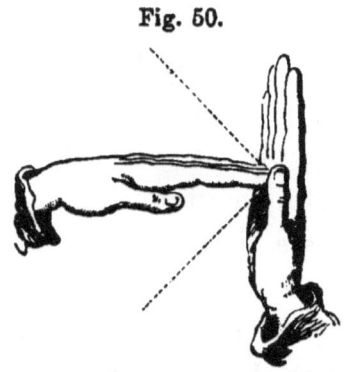

Position of hands so as to form right angles.

8

inclined beds bisect the right angle so as to give an angle of 45°, or whether the inclination be greater or less than that amount, and how much.

When strata dip in opposite directions, they form an anticlinal axis, as at A, Fig. 51; and when strata dip towards a given line, from opposite directions, they are said to form a synclinal axis, as at B, Fig. 51.

Fig. 51.

Showing Anticlinal and Synclinal Axis.

Strata sometimes appear horizontal, when highly inclined. This is possible when a vertical wall of the edges of inclined strata is presented, as is often the case on the sea-coast. Instances of this kind may be observed in deep gorges in the mountains. An examination of the strata at several points in the immediate neighborhood will prevent deception.

The *outcrop* of strata is their appearance at the surface of the earth.

Strata are *conformable* when the planes of stratification are parallel, no matter how much inclined or bent. Thus the strata represented by Figs. 48, 49, and 51 are conformable. Unconformable strata are those whose planes of stratification do not conform to the planes of those strata upon which they rest. In Fig. 52, the horizontal strata are unconformable to the inclined strata upon which they rest.

Fig. 52.

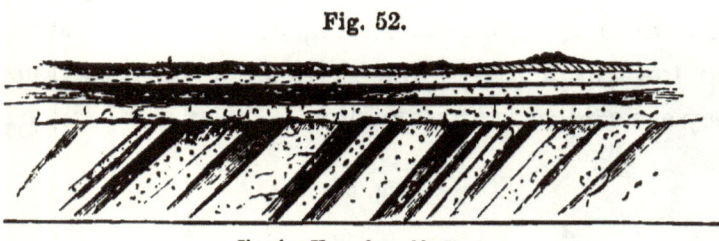

Showing Unconformable Strata.

Faults are produced by the breaking of the beds across their planes of stratification, and thus permitting the strata to slide up or down, so that the two parts of a given bed are no longer on the same level.

Fig. 53.

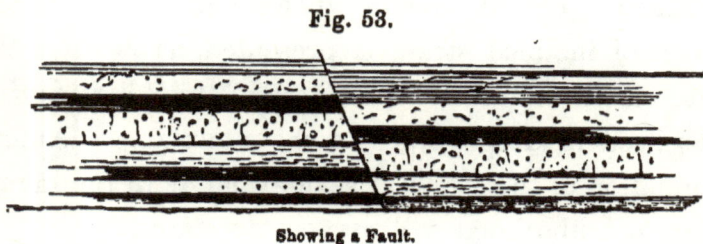

Showing a Fault.

Faults seriously retard mining operations; for suddenly the end of the bed of coal, or other substance, is reached, and the workmen know not whether its counterpart is above or below the level upon which they have been operating.

Strata which are really lowest or oldest may appear

88 STRATIFIED ROCKS.

highest. In ascending a mountain, we may constantly be reaching lower and lower strata. By examining the stratified rocks represented in Fig. 39, it will be seen that all the inclined beds were formed and elevated before the horizontal ones could accumulate upon them. It will also be seen that the lowest of the inclined beds do not make their appearance at the surface till near the summit of the mountain.

The same strata may be observed in many places, and be mistaken for others. Such a mistake is the more likely to be made, because the same beds often dip in opposite directions in the different places where observations are made. Fig. 54 illustrates a case of this kind.

Fig. 54.

Showing the same beds dipping in opposite directions.

Vertical and highly inclined strata are often, and perhaps always, parts of vast folds, whose upper portions have been removed. Thus the strata whose upturned edges are exhibited in Fig. 55, were once extended, as indicated by the dotted lines.

Strata are sometimes folded so as to bring the upper or newer beneath the lower or older, as illustrated in

STRATIFIED ROCKS.

Fig. 55.

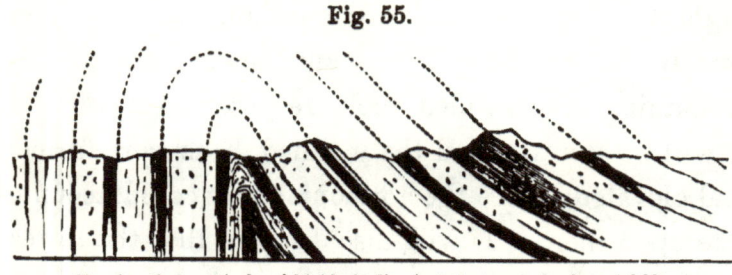

Showing that vertical and highly inclined strata are parts of great folds.

Fig 56, where the corresponding parts of a given bed are designated by the same letter.

Fig. 56.

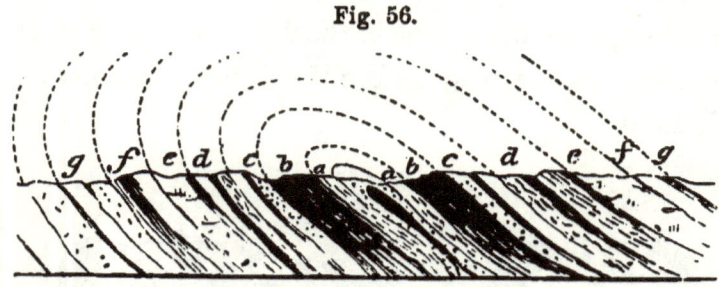

Strata folded so as to bring the newest beneath the oldest.

The various positions of the stratified rocks have resulted mainly, if not wholly, from the vertical movements which the crust of the earth has undergone from time to time. The great cause of all these movements is the molten condition of the interior of the globe. In some cases the strata have been uplifted by the force of the internal fires; but it is probable that the cooling of the heated nucleus, and the consequent contractions, have in a great measure produced the

8 *

vast folds of strata which form the hills and mountains.

The disturbances of the strata pointed out in the previous pages, show how we are able to know the condition of the rocks to a great depth; for we become as well acquainted with a series of strata by examining their upturned edges, as we should by penetrating them, were they in a horizontal position.

We are permitted to examine strata in every country, which, had they remained as originally deposited, would be miles beneath the surface. Thus the crust of the earth is laid open to actual observation to the depth of many miles.

As remarked on a previous page in regard to minerals, the student of Geology should early collect specimens of all the most common rocks, and, as far as possible, observe the various positions in which they occur.

CHAPTER VI.

GENERAL VIEW OF THE VEGETABLE AND THE ANIMAL KINGDOM, PREPARATORY TO THE STUDY OF THE REMAINS OF PLANTS AND ANIMALS IN THE ROCKS.

SECTION I.

THE VEGETABLE KINGDOM.

The Vegetable Kingdom comprises two great Branches,—Phænogamous or Flowering Plants, and Cryptogamous or Flowerless Plants.

PHÆNOGAMOUS PLANTS bear true flowers, and produce seeds having a seed-leaf or seed-leaves, called cotyledon or cotyledons, in which is enveloped a ready-formed embryo, the germ of a new plant. This branch embraces all the higher forms of vegetation, and naturally divides into two Classes,—Exogens and Endogens.

Exogens, or Outside Growers, comprise all plants whose stems are composed of three distinct parts,—pith in the centre, bark outside, and wood, or woody substance between the two. Although these parts are the most plainly exhibited in trees and shrubs, they are more or less distinctly indicated in many of the herbs, or soft-stemmed plants.

Plants of this class have net-veined leaves, and bear seeds with two or more seed-leaves, and are often called Dicotyledons. The two parts into which a bean, a pumpkin seed, or an apple seed readily divides, are the cotyledons, and they form the first two leaves of the young plant.

Exogens all grow by additions to the outside,—a new layer being added just beneath the bark each year; and thus the age of the shrub or tree is indicated by the number of concentric rings exhibited by a cross section of its stem or trunk.

Fig. 57.   Fig. 58.

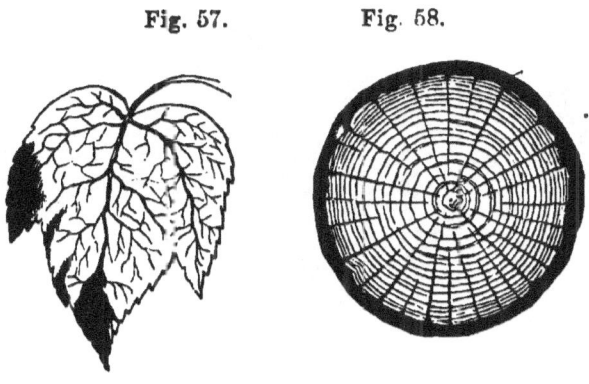

Exogenous Leaf.   Cross section of an Exogenous Stem.

As stems with concentric layers, leaves with net-like veins, and seeds with two or more cotyledons,

THE VEGETABLE KINGDOM. 93

belong exclusively to Exogens, it is plain that a stem, or a leaf, or a seed, shows at once whether it represents a plant of this class.

Exogens naturally divide into two sub-classes—Angiosperms and Gymnosperms.

Angiosperms comprise all the Exogens which bear their seeds in an ovary or seed-vessel. This is the

Fig. 59.

Cycas.

94 THE VEGETABLE KINGDOM.

case with the great body of exogenous plants,—with all except those enumerated in the following sub-class:—

Gymnosperms comprise those which bear their seeds attached to the inner surface of a scale. Such are the cone-bearing trees, as Pines, Spruces, Hemlocks, Cedars, Balsams, Larches, Yews, &c., and a family of tropical plants known under the name of *Cycas*, one species of which is represented by Fig. 59.

ENDOGENS embrace all flowering plants whose

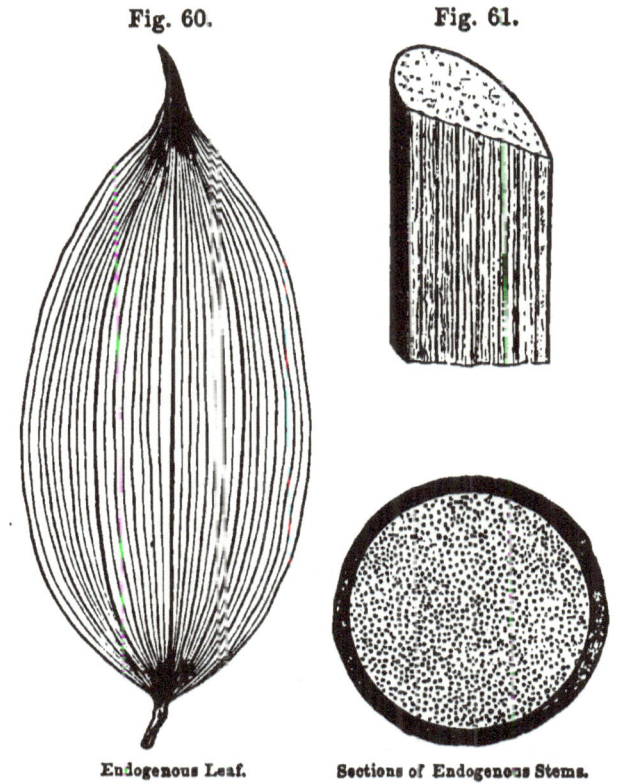

Fig. 60.   Fig. 61.

Endogenous Leaf.   Sections of Endogenous Stems.

stems are not composed of concentric layers, but whose woody substance is distributed through the stems in threads and bundles. Plants of this class have parallel-veined leaves, which sheathe the stem, and decay without falling off. They bear seeds with only one seed-leaf or cotyledon. Such are the Grasses and Grains, the Field Lilies, Solomon's Seal, Lily of the Valley, &c.; also the Palm, Sugar-Cane, Bamboo, and the like. Figs. 60 and 61 represent a leaf of the large Solomon's Seal, a cross and vertical section of a corn-stalk, and a cross section of the Palm stem.

CRYPTOGAMOUS PLANTS do not bear real flowers, nor produce seeds with a cotyledon, or cotyledons; but they bear something analogous to flowers, and produce *spores* instead of seeds. This branch comprises all the lowest forms of vegetation on the globe.

Cryptogamous Plants naturally divide into three Classes — Acrogens, Anophytes, and Thallophytes, ranking in the order in which they are named, the first being highest.

Acrogens comprise those plants whose growth is wholly or mainly at their summit. This class embraces Ferns, Club-Mosses, and Equisetaceæ, or Horsetails.

Ferns are familiar to every one under the popular name of Brakes. There are about 50 species in the Northern States. A very common form is represented by Fig 62. In most cases the spores are borne in little capsules on the back of the leaf, giving it a very beautiful appearance.

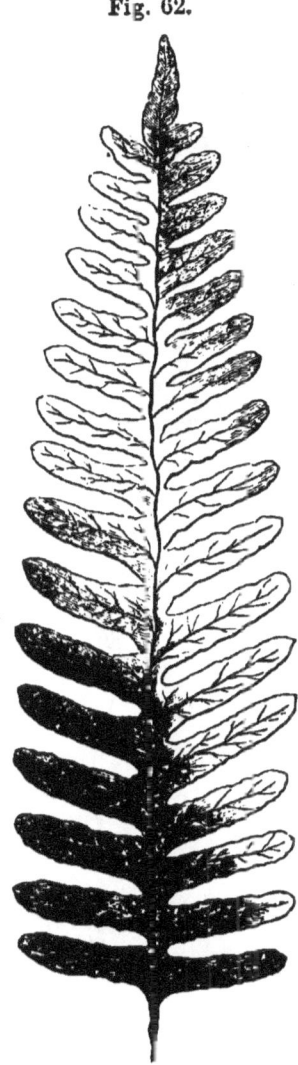

Fig. 62.

Polypedium.

In the tropics, some species of ferns grow to the proportions of trees, as in Fig. 63. In some instances they are 30 or 40 feet, or more, in height.

Club-Mosses are familiar to every one, as the leafy, evergreen, trailing plants, common in pastures and woodlands, and often gathered for festoons to adorn our houses, school-rooms, and churches. Some species send up little branches from the creeping stem, which expand into miniature trees. One species appears precisely like a little tree, and hence is called the Tree Club-Moss.

Equisetaceæ, or Horsetails, are familiar as they appear in those brown, leafless, striated, jointed stems, 6 or 8 inches long, which are

THE VEGETABLE KINGDOM.

Fig. 63.

Tree Ferns.

Fig. 64.

Equisetum.

Fig. 65.

Polytrichium.

found everywhere in early Spring; and in those similar stems, 2 or 3 feet high, which grow near streams, and contain so much silica, that they are used for scouring, and hence called Scouring Rushes. Fig. 64 gives a good idea of the plants of this group.

ANOPHYTES comprise all the true mosses.

Mosses are small plants with leafy stems, and simple narrow leaves. Fig. 65 shows one of the most common forms. The soft mosses which are so common in wet meadows, are known under the name of Sphagnum.

THALLOPHYTES comprise the Algæ,—such as seaweeds and the like,—Lichens, and Fungi. The *Algæ* exhibit a great variety of interesting forms and colors; and are very beautiful. *Lichens* occur in the form of incrustations and foliaceous expansions on rocks, trees, and fences, and, in many places, they cover the trees, in long, hanging masses, composed of innumerable threads. This last form of the lichen is most commonly found on trees growing in low lands.

Fungi are the toadstools, and the like. The leathery forms of vegetation which grow upon dead trees, and upon old logs, belong to this group.

The classification of plants may be presented in one view as seen below.

VEGETABLE KINGDOM.	PHÆNOGAMOUS, OR FLOWERING PLANTS.	EXOGENS, OR DICOTYLEDONS.	*Angiosperms.*
			Gymnosperms.
		ENDOGENS, OR MONOCOTYLEDONS.	
	CRYPTOGAMOUS, OR FLOWERLESS PLANTS.	ACROGYNS. ANOPHYTES. THALLOPHYTES.	

A FEW WORDS ABOUT THE DISTRIBUTION OF PLANTS.

The distribution of plants is in accordance with climate; each climatic zone being characterized by a peculiar flora. But while each climate has a flora, which, for the most part, presents the same general characteristics in different countries, it is also true, that, excepting the frozen regions, the species of plants differ widely in different countries, even in the same climatic zone.

It is in the Torrid Zone that vegetation reaches its highest expression, both in number of species, and in luxuriance of growth.

From the tropics to the poles, vegetation gradually diminishes, both in species and in luxuriance; and in the frozen regions, the whole flora is composed of dwarfed shrubs, mosses, and lichens.

Vegetation changes with the increase of altitude, in the same manner as with the increase of latitude. The highest mountains in the tropics exhibit the zones of vegetation in the same order as they occur in passing from the equator to the polar regions. At the base of the mountains we find tropical luxuriance, but soon the Palm, the Bamboo, and Tree Fern, give way to the Oak and the Chestnut, and these in turn to the Pine and other Conifers; and then comes the region of dwarf Birches, and Alpine Shrubs, and higher still, are only the Mosses and Lichens of the Arctic regions.

In the Arctic Zone the same species of plants are found in North America, Greenland, Europe, and Asia; but, passing from this zone towards the south, we find the plants of the two hemispheres more and more unlike, even in similar climates, till there is the greatest diversity between the floras of South America, Africa, and Australia.

The resemblance between the flora of Europe and that of North America, leads the superficial observer to suppose them identical, which is not the case. We have but few plants identical with those of Europe.

In North America, there are at least 150 species of Forest Trees, while in Europe there are less than 40 species. Moreover, the Pines, Elms, Oaks, Chestnuts, and Lindens of Europe, are specifically different from the trees bearing the same names in North America.

There are probably 150,000 species of plants now upon the earth. It is worthy of note, that no country, however diversified, nor hemisphere even, can furnish them all. Nothing short of the whole earth can present us all the species of the Vegetable Kingdom. This great number of species is distributed among many botanical provinces, each of which has a flora found nowhere else.

Plants are distributed through the agency of man and animals, especially of birds; and through the agency of winds, rivers, waves, and currents. By these means the species of the different regions, to a limited extent, have been interchanged. But the distribution of the plants into great botanical provinces, resulted from no such accidental causes; nor is this distribution the result of the influence of climate, though in accordance with it.

The above and similar facts show that the vegetation of the globe has not been distributed from any one country. On the contrary, the species, for the most part, have originated near those great centres around which we now find them clustered. As fast as the lands rose from the water, they were clothed with such vegetation as they were adapted to support; but all done according to a Great Plan made by "Him who ruleth over all."

SECTION II.

THE ANIMAL KINGDOM.

There are not less than 250,000 species of animals living on the earth. To study these in all their various forms and relations, is the delightful task of the zoölogist. But it is absolutely necessary that the student who would investigate the history of our globe, should become acquainted with the leading facts of the Animal Kingdom; for it is only by knowing something of animal life, as it now appears on the earth, that he can have any true understanding of the remains of animals found in the rocks.

Our brief limits allow only the most general statement of results which have been reached, after long years of toil, by such men as Cuvier, Agassiz, and other great and noble minds. I would invite the attention of the student to their works, and would especially urge the study of animals, so far as circumstances will permit.

The natural divisions of the Animal Kingdom, are *Branches*, or *Types*, *Classes*, *Orders*, *Families*, *Genera*, and *Species*. That is, the animal kingdom is divided into branches; each branch into classes; each class into orders; each order into families; each family into genera; and each genus into species composed of individuals which are essentially alike.

Agassiz has shown that these divisions are not the invention of man for his own convenience, but that they exist in nature. He has also shown that

"*Branches*, or *Types*, are characterized by the plan of structure.

Classes, by the manner in which that plan is executed, as far as ways and means are concerned.

Orders, by the degrees of complication of that structure.

Families, by form.

Genera, by the details of execution in special parts.

Species, by the relations of individuals to one another, and to the world in which they live, as well as by the proportion of their parts, their ornamentation, &c."

The Animal Kingdom consists of four great branches, or types—Radiates, Molluscs, Articulates, and Vertebrates. All the animals in any one of these branches are built upon the same plan, or, in other words, after the same type.

RADIATES comprise all animals whose organs radiate from a common point or centre. There are probably 10,000 living species. This branch contains three Classes—Polyps, Acalephs, and Echinoderms.

POLYPS comprise animals with a fleshy, tubular, or sack-like body, having a circular summit or disk, in the centre of which is the mouth, surrounded by one or more rows of tentacles. They are all marine, and are attached, by their lower extremity, to submarine bodies, and to the sea bottom. They abound on the coasts in shallow water, and, with few exceptions, do not live deeper than 20 or 30 fathoms. Within these limits different species flourish at different depths.

From their resemblance to plants, they are often called Zoöphytes. They vary in size, from a microscopic point to eighteen inches in diameter. They increase by eggs, by budding similarly to plants, and by divisions and subdivisions. The polyps known as Sea-Anemone, or Actinia, so common on our own coast, give the student a perfect idea of the animals of this class.

Polyps are both solitary and associated, frequently in the most astonishing numbers. The largest community, however, is only the increase of a single individual.

Some polyps secrete calcareous matter. In this respect, there is a gradual passage from the actinias, all of which are fleshy, to those polyps which secrete a solid framework, or *Coral*. Yet there is no essential difference in external appearance, or structure, between the actinias and the coral-producing polyps, excepting only the fact that the former are generally larger than the latter.

The too common impression that polyps build

Fig. 66.

Expanded.

Fig. 67. Fig. 68.

The same closed.   *The same opening.*

Sea-Anemone, or Actinia, Coast of Massachusetts.

coral at will, as the bee builds comb, or as workmen masonry, is entirely erroneous. Coral is simply the skeleton or aggregate skeletons of polyps, and is a necessary result of their existence. In fact, the polyps form coral in the same manner as the higher animals form bones.

Every part of live coral is wholly enveloped by the polyps that produced it, and in no sense are the polyps within the coral, though when disturbed they contract, and thus give the appearance of retreating within it.

Coral animals are not always minute, but on the contrary often a quarter or half an inch, and sometimes several inches, in diameter.

Each of the cells on a piece of coral shows the position occupied by the coral animal; and hence by counting the cells, we learn the number of polyps engaged in secreting a particular mass. The size of the depression also shows the size of the animal.

Fig. 69.

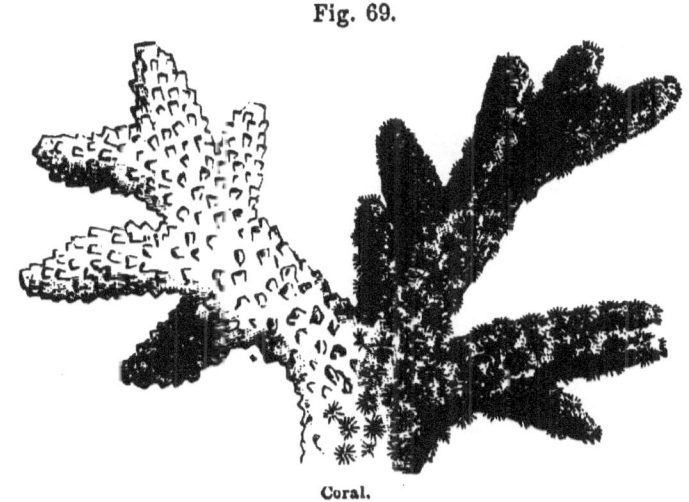

Coral.

Fig. 69 represents one of the common forms of coral, called Madrepore. The branch on the right is represented as alive, the other as dead.

The forms and hues of live coral are almost endless. Some parts of the tropical seas, where polyps especially flourish, rival in graceful and varied forms, and in splendor of colors, the most beautiful flower-gardens of the lands. There is scarcely a form of vegetation, either trunk or branch, leaf or flower, fern, moss, lichen or fungus, that is not imitated to the life by these wonderful animals of the sea, whose united skeletons at length form islands which in due time become the home of man.

Polyps may be divided into two Orders —Actinoids, embracing all the Actinias, and allied coral polyps; and Alcyonoids, embracing polyps having only eight tentacles.

For a perfect exposition of the whole subject of Polyps, the student is referred to Professor J. D. Dana's splendid work on Zoöphytes.

Sponges, formerly regarded as Zoöphytes, are now by many eminent naturalists, considered as belonging to the vegetable kingdom.

ACALEPHS, so called on account of their irritating properties, embrace all the Jelly-fishes, or Medusæ, and their allies.

Scientifically considered, Acalephs comprise three Orders—Hydroids, formerly regarded as polyps, Discophoræ, and Ctenophoræ.

108 THE ANIMAL KINGDOM.

Hydroids comprise little plant-like marine animals, and the so called fresh-water polyps. Fig. 70 represents the Hydra, the well known fresh-water polyp, found in brooks and ponds. It is attached at base, and its long tentacles are spread to entrap some little animal for food. Fig. 71 shows a marine hydroid magnified, called Coryne.

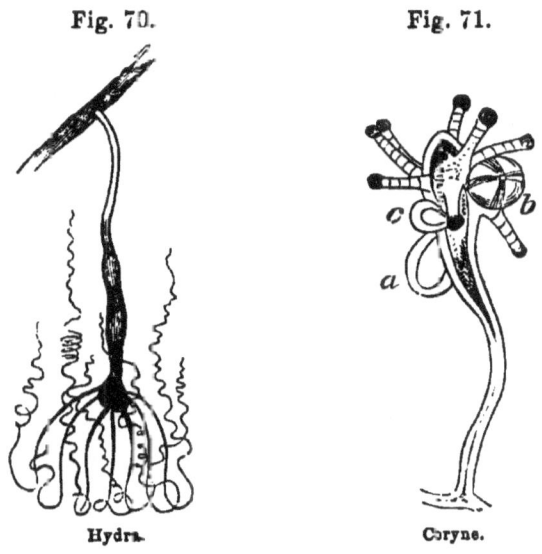

Fig. 70. Fig. 71.

Hydra. Coryne.

Discophoræ comprise the disc, and bell-shaped Medusæ. In fact, marine Hydroids and the Discophoræ are the same in different stages of development; or, in other words, the Medusæ of this order are born of marine hydroids; and hence, Agassiz, I believe, groups them both under the term *Hydro-Medusæ*. Fig. 71, copied by permission from the splendid plates for the forthcoming volume on this subject, by Agassiz, shows the buds *a, b,* of the hydroid just swelling into free jelly-fishes, like that represented by Fig. 72. Another, in a less advanced stage, is seen at *c*. The hydro-medusæ are at present attracting much attention from the ablest naturalists.

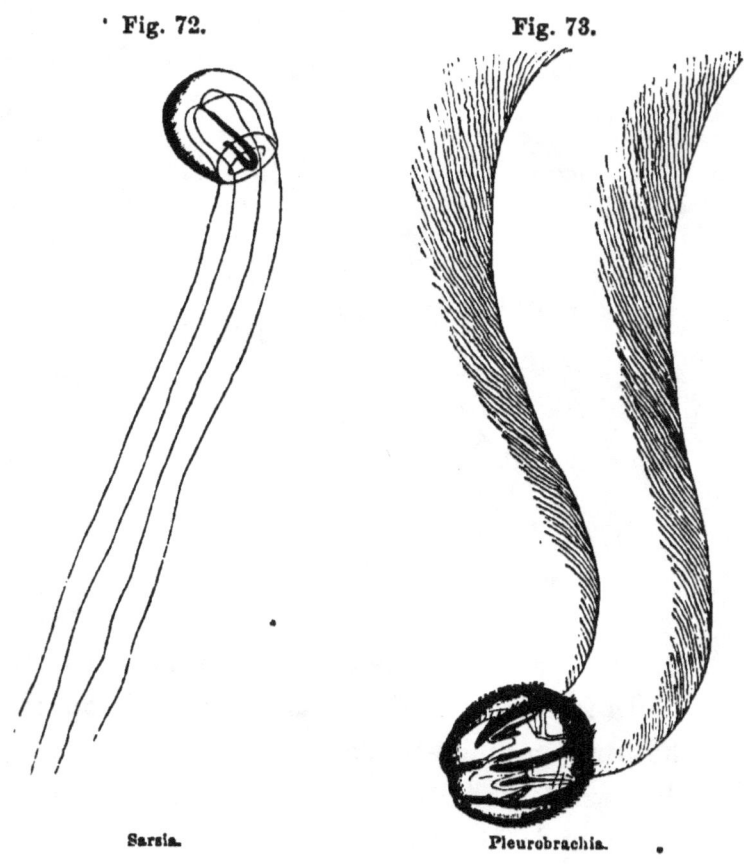

Fig. 72. Fig. 73.

Sarsia. Pleurobrachia.

In one of his papers, Agassiz has remarked as follows, about these curious animals of the sea:—"It is in reality one of the most wonderful sights which the philosophic naturalist can behold, to see animals scarcely more dense than the water in which they play, and almost as limpid, perform, in that medium, movements as varied as those of the eagle which soars in the air, or the butterfly dancing from flower to flower, testifying, by their activity, their sensitiveness and their volition."

The Jelly-fishes here alluded to are small. Some other species grow to a great size, weighing, sometimes, 30 or 40 pounds; and these, too, are born of hydroids of different kinds. One of the most common species is the umbrella-shaped one, seen everywhere along our coast, popularly known as the Sunfish.

Ctenophoræ are better represented by Fig. 73, than by any description that I can give.

ECHINODERMS comprise marine animals which mostly have a calcareous covering bearing spines, though one order is made up of animals which have a thick tough skin.

This class embraces four Orders,—Crinoids, Asteroids, Echinoids, and Holothurioids.

Crinoids are lily-shaped animals, which are attached by a sort of stem to the sea bottom. They resemble some forms of vegetation, but are, nevertheless, real animals, and are closely related to the Asteroids spoken of on the next page. In fact, they appear like a sort of Star-fish with a stem; yet it must be borne in mind that there is no such relationship as parent and offspring between crinoids and star-fishes. There is only one species of pedunculated crinoid now

Fig. 74.

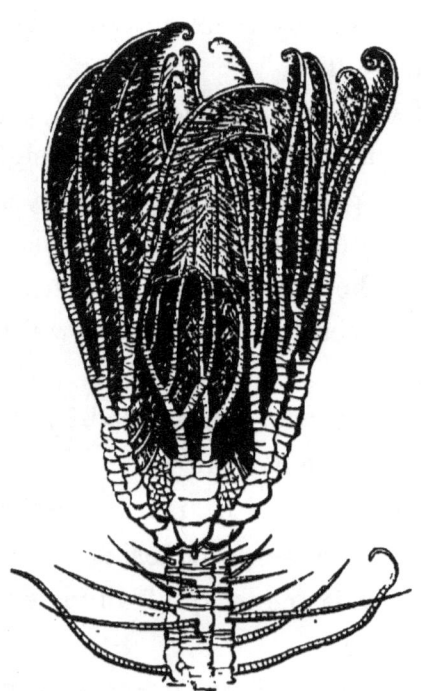

Pentacrinus Caput-Medusæ, West Indies.

living, the *Pentacrinus Caput-Medusæ* of the West Indies, represented by Fig. 74. We shall find that they are very abundant in the rocks.

Asteroids comprise the Star-fishes, which are com-

112 THE ANIMAL KINGDOM.

mon everywhere on our rocky coasts. They are readily found at low tide, by turning up the sea-weed and looking in the clefts of the rocks.

Fig. 75.

Star-fish, Nahant, Mass.

Echinoids include Sea-urchins, Spatangoids, and the like. Figs. 76 and 77 give the student a good idea of the animals of this order. Though at first view, Echinoids appear very different from Star-fishes, a little patient study will show that all the parts of the one, correspond to those of the other,—in a word, that both are built upon the same plan. It will afford the student one of the most pleasant and profitable exercises to trace out the homology of the star-fish and sea-urchins.

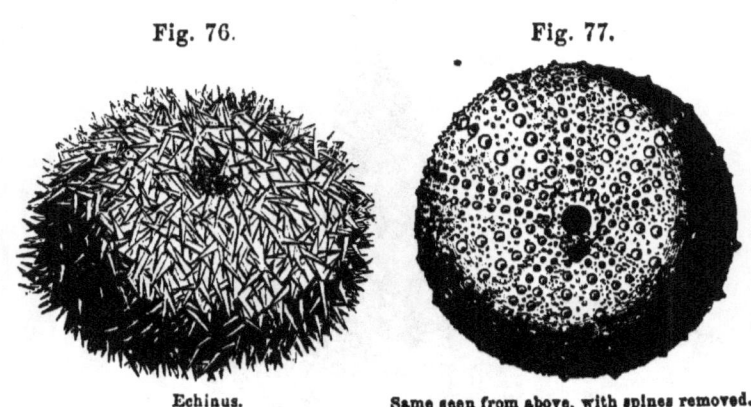

Fig. 76. Fig. 77.

Echinus.   Same seen from above, with spines removed.

Holothurioids include animals with an elongated, worm-like body, and a leathery skin. Such are the Sea-Slugs.

MOLLUSCS are built upon a plan entirely different from the Radiates. They are all soft-bodied animals. The shell found as a covering in many species, may be compared to an external skeleton. They are terrestrial, fresh-water, and marine. Number of living species 16,000. This branch comprises three Classes —Acephals, Gasteropods, and Cephalopods.

ACEPHALS embrace four Orders—Bryozoa, Brachiopods, Tunicata, and Lamellibranchiates.

Bryozoa, also called Moss Animalcules, are small

molluscs growing in clusters, forming incrustations on rocks, and other submarine bodies. They resemble corals.

Brachiopods comprise those bivalve molluscs whose two valves are never equal, but are always equal-sided. From the position which the animal occupies, the two valves of the shell are called respectively dorsal and ventral. The ventral valve is largest, and has a prominent beak through which the organ of adhesion passes; for brachiopods grow attached to submarine bodies. The dorsal valve is always free and imperforate.

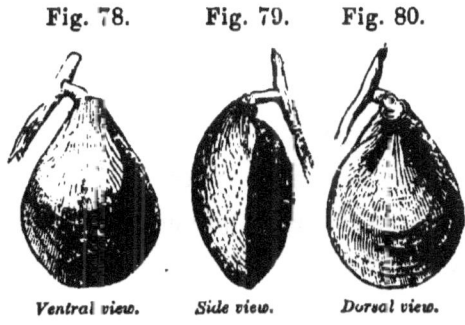

Fig. 78. Fig. 79. Fig. 80.

Ventral view.   *Side view.*   *Dorsal view.*

Terebratula, a Brachiopod, Coast of Maine.

Tunicata are those molluscs that are protected by an elastic covering or tunic instead of a shell.

Lamellibranchiates are those molluscs which have their gills in lamellæ. Such are the oyster, marine and fresh-water clams, and the like.

GASTEROPODS embrace three Orders—Pteropods, Heteropods, and Gasteropods proper.

Pteropods are small oceanic snails with wing-like appendages. They are the food of the right whale.

Heteropods comprise oceanic snails with a very tender shell.

Gasteropods proper are such as effect locomotion on a fleshy foot. Such are our common land-snails, also the Natica, and the like on our coasts.

A group of small animals, mostly microscopic, called Foraminifera, is placed under this order by some modern writers, though by others not considered as strictly belonging to any of the four great branches established by Cuvier. These animals have a calcareous shell, often in the form of a snail shell, with many holes through which they extend feelers, with which they catch their prey. We shall learn hereafter that these animals are abundant in the rocks.

CEPHALOPODS comprise marine molluscs, whose principal appendages are attached to the head. Some of them have fins, and all can propel themselves by the forcible expulsion of water from a cavity, or chamber, with which they are provided. This class is divided into two Orders—Tetrabranchiates and Dibranchiates.

Tetrabranchiates are cephalopods which breathe by four gills. They have an external shell, divided into partitions or chambers, which are connected by a tube or siphuncle. They are often called chambered-shelled molluscs. Although more than 1400 species

Fig. 81.

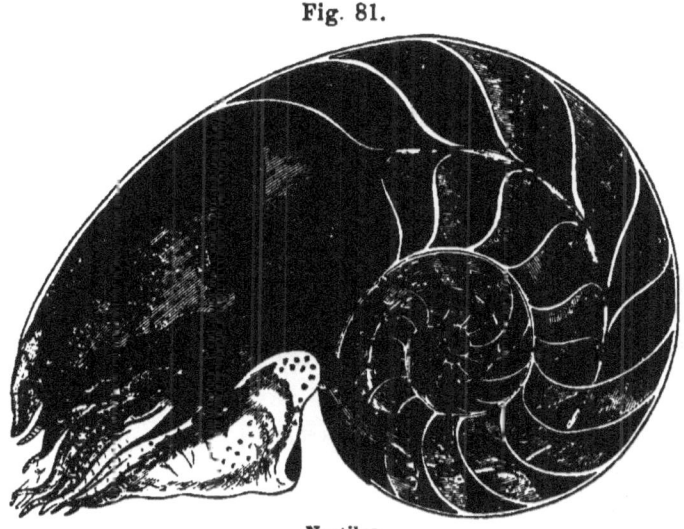

Nautilus.

have been found in the rocks, the Nautilus is the only living representative. Fig. 81 represents the Nautilus cut open, showing the animal lying in the shell; also the position of the chambers and siphuncle. The chambers are formed as the animal grows, and moves forward, leaving one partition after another.

Dibranchiates are those cephalopods which breathe by two gills. The common Cuttle-fish and Squids are familiar examples. With the exception of a single genus they are naked—having their solid portion, a sort of rudimentary shell, inside. They have eight or ten muscular arms and powerful jaws. They are provided with a sack or ink-bag, from which, when pursued, they discharge an inky fluid, rendering the water turbid, and thereby escape.

Fig. 82 gives a good idea of the animals of this order. The Argonauta, or Paper Sailor, is a representative of the only genus which has an external shell.

Fig. 82.

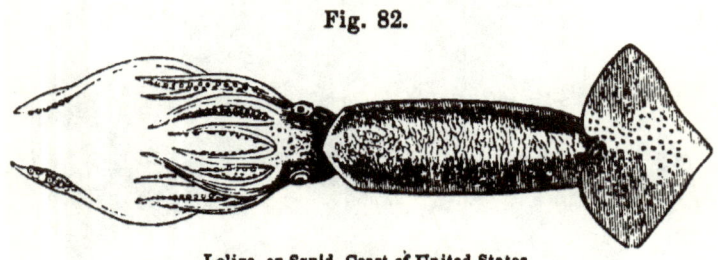

Loligo, or Squid, Coast of United States.

ARTICULATES comprise those animals which have the body more or less divided into lobes, rings, or joints, often movable upon one another. Their hard skin forms a sort of external skeleton, the muscles being attached to the inside. Estimated number of living species 200,000. This branch may be divided into three natural Classes—Worms, Crustaceans, and Insects.

WORMS comprehend Helminths, or Intestinal Worms, Earthworms, Leeches, &c.

CRUSTACEANS are divided into four Orders—Rotifera, Entomostraca, Tetradecapods, and Decapods.

Rotifera are all microscopic animals.

Entomostraca include the Limulus, or Horse-shoe Crab, and the Cirripeds, that is the Barnacles. Besides the barnacles, represented by Fig. 84, there are a great many species with a sub-conical shell, which everywhere cover the rocks that are daily washed by the tide.

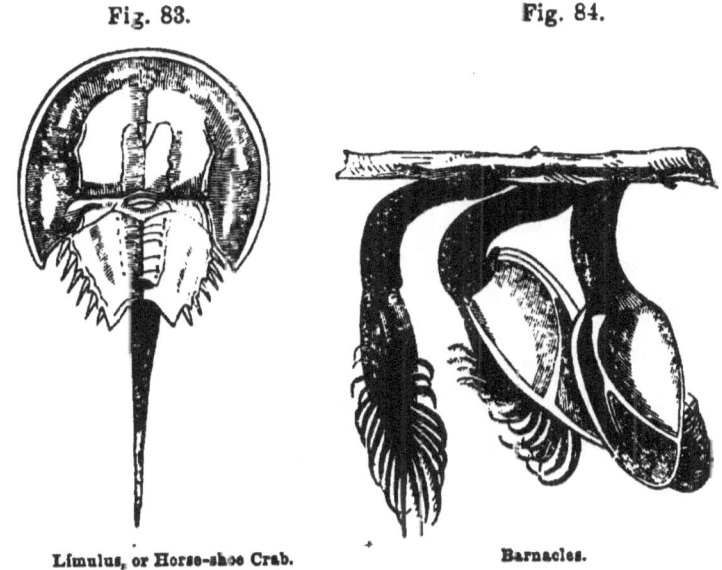

Fig. 83. Limulus, or Horse-shoe Crab.

Fig. 84. Barnacles.

Tetradecapods are the Sand-fleas and the like.

Decapods comprise Lobsters, Crabs, Shrimps, and their allies.

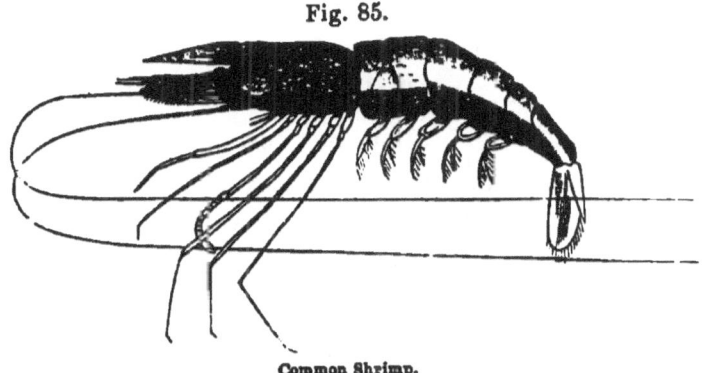

Fig. 85. Common Shrimp.

Insects contain three Orders, according to Agassiz—Myriopods, Arachnids, and Insects proper.

Myriopods comprise the Millipedes and Centipedes.

Arachnids comprise the Spiders and Scorpions.

Insects proper comprise Beetles, Bugs, Grasshoppers, Darning-needles, Bees and Wasps, Aphides or Plant-lice, Harvest-fly, Butterflies, House-flies, and Musquitoes, represented by thousands of species. This order contains more species than any other in the animal kingdom.

VERTEBRATES embrace all animals which have two cavities, more or less elongated, one above, and the other below a bony or cartilaginous axis called the back bone. The upper cavity contains the spinal cord, which, at one extremity, is enlarged into a lobe, or lobes called the brain. The lower cavity contains the organs of respiration, digestion, and reproduction.

In all vertebrates the skeleton is internal, and constitutes the frame upon which the muscles are placed; the skin, with its appendages, surrounding the whole. In most of the vertebrates, the axis of the skeleton is made up of parts, called vertebræ; which are more or less movable one upon another.

This branch contains five natural Classes,—Fishes, Batrachians, Reptiles, Birds, and Mammals.

That the animals in all these classes are built upon the same plan, may be seen by a little careful study. They all exhibit the two cavities, before mentioned, one above and the other below the main axis of the body. They all have an internal skeleton. The unity of the plan upon which all vertebrates are built,

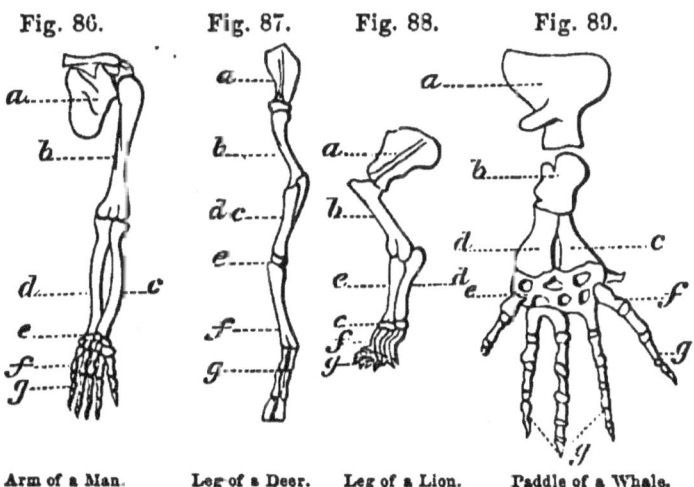

Arm of a Man.   Leg of a Deer.   Leg of a Lion.   Paddle of a Whale.

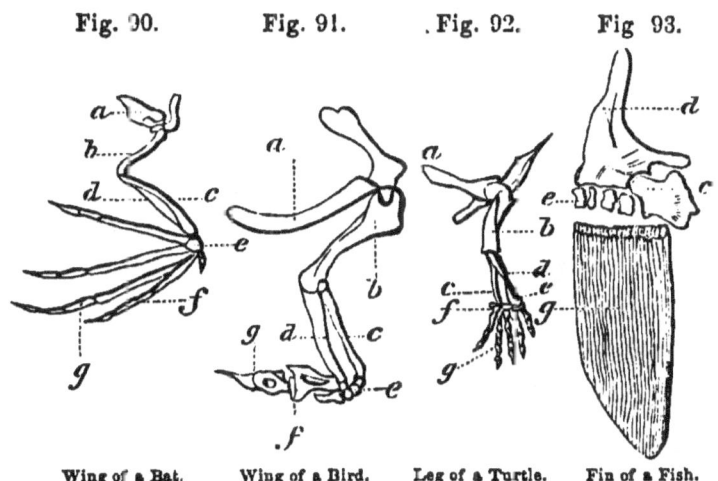

Wing of a Bat.   Wing of a Bird.   Leg of a Turtle.   Fin of a Fish.

is strikingly brought out by a comparison of their skeletons, or even of their anterior members of locomotion. The foregoing representations of the forward locomotive members of vertebrates, in which corresponding parts are designated by the same letter, show that they are all one and the same thing, expressed in different ways.

FISHES may be divided into four Orders, if we adopt the earlier writings of Agassiz—Placoids, Ganoids, Ctenoids, and Cycloids; this classification being based upon the shape of the scale, which is entirely different in the different orders. The accompanying figures give a general idea of the shape of the scales in each order. The number of species of living fishes amounts to about 10,000.

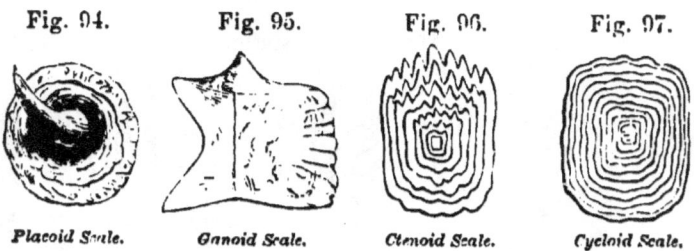

Fig. 94.   Fig. 95.   Fig. 96.   Fig. 97.

Placoid Scale.   *Ganoid Scale.*   *Ctenoid Scale.*   *Cycloid Scale.*

Form of the Scales in the different Orders of Fishes.

Placoids comprise fishes with flat scales and a cartilaginous skeleton. Such are the Sharks, and the Scates. The scales of the latter are armed with a sharp curved spine.

Ganoids comprise fishes which have enamelled scales. Such are the Gar-pike, and Sturgeon.

Placoids and Ganoids are characterized by unequal lobed or heterocercal tails, the spinal column being prolonged into the upper lobe. These are the two lowest orders of fishes. The two following orders have a single tail fin, or the tail is equally bilobate or homocercal.

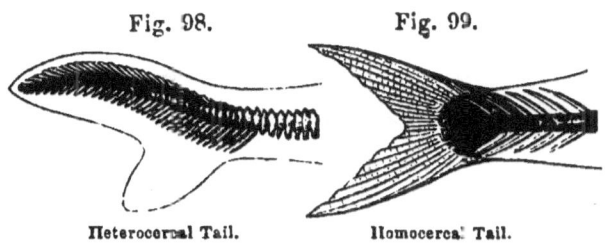

Fig. 98. Heterocercal Tail.   Fig. 99. Homocercal Tail.

Ctenoids embrace all fishes which have scales toothed on the edge. Such are the Perch, Bream, Bass, Flounder, &c.

Cycloids are those fishes whose scales are rounded and entire. Such are the Salmon, Cod, Mackerel, Pickerel, Sucker, Trout, &c.

The last two orders comprise all those fishes which are the most useful for the food of man.

BATRACHIANS embrace three Orders—Frogs, Salamanders, and Cæcilians, or Snake-like Batrachians.

REPTILES comprise four Orders—Rhizodonts, Saurians, Chelonians, and Ophidians. The animals of this class, as well as those of the two preceding ones, are cold-blooded, or rather the temperature of their blood changes with that of the medium in which they live. Reptiles and Batrachians are represented by about 2000 living species.

Rhizodonts are large reptiles with hollow teeth. This order has but few living representatives.

Saurians embrace the Crocodile, Alligator, Lizards, and Glass-snakes.

Chelonians embrace all Turtles or Tortoises, whether land, fresh-water, or marine.

Ophidians comprehend all the snake families.

BIRDS may be divided into seven Orders—Natatores, Grallæ, Cursores, Rasores, Scansores, Insessores, Raptores. The first two orders embrace all the aquatic birds. Birds have warm blood. Between 6000 and 7000 species are known.

Natatores, or Swimming Birds, are those which have rather short legs, webbed feet, and other peculiar adaptations to the water, which is their principal place of resort. Such are Ducks, Mergansers, Loons, Petrels, Gulls, Pelicans, &c.

Grallæ, or Waders, embrace those birds which frequent the water, wading in the shallows in search of food. They are characterized by having legs, neck, and bill all very long. Such are Herons, Plovers, Snipes, Rails; also the Flamingo, which, however, by its bill and webbed feet, seems to form a transition between this order and the last.

Cursores, or Runners, are the Ostriches. There are only five species now living—one in Africa, one in India, one in Australia, and two in South America.

Rasores, or Scratchers, comprise the Turkey, Hen, Grouse, Quails, and their allies. They are all well adapted to live principally on the ground. Doves, the young of which hatch in a very feeble condition, form a sort of transition between the Insessores and Rasores.

Scansores, or Climbers, comprise the Woodpeckers and Parrots. Having two toes turned forward, and two behind, they are particularly adapted to climbing along the trees in search of food.

Insessores, or Perchers, often called Oscines, comprise our most common birds, such as Crows, Jays, Thrushes, Finches, Warblers, and their hundreds of allies.

Raptores, or Raveners, comprise the Eagles, Hawks, Owls, Vultures, and the like. The birds of this order are characterized by powerful muscles, sharp claws, and strong hooked bills, all of which, together with their great extent of wing, aid them in capturing birds and other animals for food. With the exception of Vultures, that live on dead animals which they chance to find, all the birds of this order catch their own prey.

MAMMALS comprise all those animals which bring forth their young alive, and nourish them from their own bodies. They all have warm blood. Mammals are represented by about 2000 living species. Including Man, the highest representative of this class, Mammals comprise 10 Orders—Marsupials, Pachyderms, Cetaceans, Edentata, Insectivora, Rodents, Ruminants, Carnivora, Quadrumana, and Bimana.

Marsupials, or Pouched animals, comprise the Opossum, Kangaroo, &c. All the mammals of Australia belong to this order.

Pachyderms, or Thick-skinned animals, include the Horse, Hippopotamus, Rhinoceros, Elephant, Hog, Tapir, &c.

Cetaceans comprise the Whales, Porpoises, Dolphins, and the Narwhal.

Edentata are Toothless animals; such as Sloths, Ant-eater, and Armadillo.

Insectivora, or Insect-eating animals, include the Bat, Mole, Hedgehog, &c.

Rodents, or Gnawers, include the Hare, Squirrel, Beaver, &c.

Ruminants include the Ox, Sheep, Goat, Giraffe Deer, Moose, &c.

Carnivora, or Flesh-eaters, include the Cat tribe, Dog, Otter, Mink, Seal, Bear, Badger, &c.

Quadrumana, or Four-handed, comprise all Monkeys. *Bimana*, or Man.

The Classification of the Animal Kingdom, as far as Orders, may be presented in one view, as seen on the next page.

THE ANIMAL KINGDOM.

ANIMAL KINGDOM.
- VERTEBRATES.
  - MAMMALS.
    - Bimana.
    - Quadrumana.
    - Carnivora.
    - Ruminants.
    - Rodents.
    - Insectivora.
    - Edentata.
    - Cetaceans.
    - Pachyderms.
    - Marsupials.
  - BIRDS.
    - Raptores.
    - Insessores.
    - Scansores.
    - Rasores.
    - Cursores.
    - Grallæ.
    - Natatores.
  - REPTILES.
    - Ophidians.
    - Chelonians.
    - Saurians.
    - Rhizodonts.
  - BATRA'NS.
    - Frogs.
    - Salamanders.
    - Cæcilians.
  - FISHES.
    - Cycloids.
    - Ctenoids.
    - Ganoids.
    - Placoids.
- ARTICULATES.
  - INSECTS.
    - Insects proper.
    - Arachnids.
    - Myriopods.
  - CRUSTA'NS.
    - Decapods.
    - Tetradecapods.
    - Entomostraca.
    - Rotifera.
  - WORMS.
    - Anellides.
    - Nematoids.
    - Trematods.

ANIMAL KINGDOM.—(Continued.)
- MOLLUSCS.
  - CEPHALOPODS.
    - Dibranchiates.
    - Tetrabranchiates.
  - GASTEROPODS.
    - Gasteropods proper.
    - Heteropods.
    - Pteropods.
  - ACEPHALS.
    - Lamellibranchiates.
    - Tunicata.
    - Brachiopods.
    - Bryozoa.
- RADIATES.
  - ECHINODERMS.
    - Holothurioids.
    - Echinoids.
    - Asteroids.
    - Crinoids.
  - ACALEPHS.
    - Ctenophoræ.
    - Discophoræ.
    - Hydroids.
  - POLYPS.
    - Actinoids.
    - Alcyonoids.

Note.—Besides the animals already noticed, there are innumerable microscopic organisms in all kinds of water, which Ehrenberg considered as animals, and which he grouped under the general term Infusoria. Later researches have shown that many of these are algæ; and now all these lowest forms of life are described by many writers under the name Protozoa. Many of these organisms have a silicious shell, and they form extensive layers on the bottom of our ponds, consisting wholly of their silicious skeletons, which, under the microscope, appear like small glass boxes.

A FEW WORDS ABOUT THE DISTRIBUTION OF ANIMALS.

We have seen in the last section that vegetation differs in different climatic zones, and in different geographical regions. The same is true of animals. Each climatic zone, and each grand division of the earth's surface, has animals that are peculiar to itself. Not only do the same climates have different animals in the different continents, but often in different parts of the same continent.

As in the case of plants, it is in the torrid zone that animal life reaches its highest expression; excepting only the marine mammals, which exhibit their highest forms in the Whale, the Walrus, and the Seal of the frigid regions.

In the Arctic zone the same species inhabit all the

countries within its borders. But, as remarked in regard to plants, the animals of the two hemispheres have less and less of specific identity, as we pass from this zone towards the south, until we find not even a family resemblance between the animals of Australia, Southern Africa, and the southern part of South America.

It is true that the animals of different continents, in the same climates, generally resemble each other; but in nearly all cases, except in the frigid zone as mentioned above, it is mere resemblance, and not specific identity. A few facts will illustrate these statements.

The animals of Europe and the United States resemble each other so closely that the early settlers of this country applied the names of the species they had known in Europe to the similar American species —an oversight which has caused confusion in names ever since; for we have but few species of animals identical with those of Europe. The Bears, the Wolves, the Foxes, the Wild-cats, the Deer, the Beaver, the Squirrels, the Hares, and the Birds of America, though resembling those of Europe and Asia, are not the same species; and the marks by which they can be distinguished from one another are readily found by the experienced naturalist.

Again, there are 91 species of Monkeys in America, and 79 species in the Old World; yet there is not one species, nor genus, nor family even, common to the two hemispheres. The monkeys of the New World are characterized from those of the Old by their nostrils being wide apart; by having three false grinders on each side of both jaws; by having cheek pouches; and by the prehensile tail of many species—all of which characteristics are entirely wanting in the monkeys of the Old World. The monkeys of America are smaller and less ferocious than those of the Eastern Hemisphere. Monkeys are found in nearly all the countries of the tropics, except New Guinea and Australia, and the Pacific Islands between these and the west coast of America. In the New World only one species exists on the west side of the Andes, 90 being found east of the same range. This fact shows how faunas are limited by mountain ranges.

In the tropics of the Old World the Carnivorous animals are represented by the Lion, Leopard, Tiger, &c., while in tropical America we find only the Puma, Jaguar, and the like. The tropics of the Old World furnish those huge Pachyderms, the Elephant, Rhinoceros, and Hippopotamus; while these are wanting in America, and their places are supplied with much feebler animals, the Tapir and the Peccary. But even in the Old World, the Elephant is not the same in Africa as it is in Asia; but each of these countries, and its immediate dependencies, has a species of elephant peculiar to itself. The same is true of the

Rhinoceros. This animal is not of the same species in Africa as it is in Asia; and the East Indies furnish several species different from both the African and the Asiatic. As a matter of fact, there are six distinct species of rhinoceros—probably seven—and these inhabit countries whose climates are not materially unlike.

Australia presents us with some striking facts which deserve mention here. This vast island, though lying with a large part of its northern half within the tropics, has no Monkeys, no Pachyderms, no Edentata, and no Ruminants. The mammals of this island are all Marsupials; and it is a remarkable fact, that except in Australia and vicinity, no marsupials exist in the Old World, and only one family—the Opossum—is found in America.

The Galapagos Islands exhibit facts no less striking. These islands not only differ in their flora and fauna from every other portion of the world, but each island of the group has its peculiar plants and animals—such as are not found in any of the others.

In tropical America there are 300 species of Humming-birds, while not a single species is known in any part of the Old World. A volume of similar facts might easily be recorded, but these will give a general idea of this part of our subject.

The same principles, which are indicated by the above facts, apply to the distribution of marine animals. The ocean, no less than the land, is divided into zoölogical provinces, each with its own peculiar species of animals. Each coast has animals which are found nowhere else. Of the two hundred species of molluscs living on the coast of New England, fifty are never found north of Cape Cod, and over eighty species are never found south of that Cape.

From the facts stated above, we learn that climate has no power to mould or shape the species of animals—and the same is true in regard to plants—or to change one species into another. Were it so, any given climate would produce, in the course of time, the same species of animals in all the countries within its limits. But so far from this being the case, we find, that, in spite of the influence of climate, animals of the different countries of the same climatic zone are specifically if not generically distinct, and in many cases even family resemblance is wanting.

Although through the agency of man, and in many other ways, animals of one region or country have been introduced into another, we are not to look to any such accidental operations for an explanation of the distribution of animals into many well-marked zoölogical provinces. On the contrary, the careful observer is led to believe that animals as well as plants, have been created by an Omniscient Being, in the places, and for the places, which they now occupy.

CHAPTER VII.

FOSSILS, AND CLASSIFICATION OF THE ROCK FORMATIONS.

SECTION I.

FOSSILS.

REMAINS of plants and animals are imbedded in many of the stratified rocks of every country. These remains are called *Fossils*, and are among the most important aids in making out a history of our earth. The department of Geology which treats specially of fossils, is called Paleontology.

In general, only the hard parts of animals are preserved, the soft parts having disappeared. Corals, shells, and Crinoids, are found as perfect in form as those in our present seas. The beautiful Crinoid represented on the next page, was dug out of the limestone near St. Louis, Mo. Leaves and stems of plants often show the most delicate markings; and ferns especially are found imbedded in the slates, as perfect in outline as though preserved in an herbarium.

Wood completely changed to silica shows the vegetable structure so plainly that the family of plants to which it belongs can readily be determined.

Fig. 100.

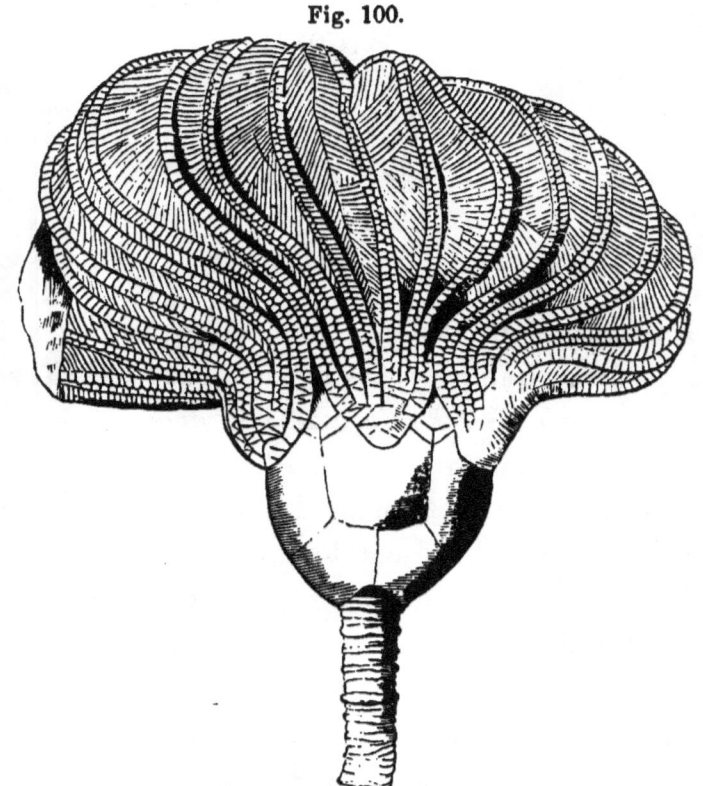

Crinoid, from the Limestone near St. Louis, Mo.

In some cases the organic body disappears, leaving an exact mould of itself impressed in the rock, and this becoming filled with mineral matter, a perfect cast of the organic body is formed. This very often happens in the case of shells, which sometimes leave a cast of the form of the outside, and sometimes of the inside. A cast of the inside is exhibited whenever the shell itself disappears after it has been filled with consolidated mineral matter.

As the organic bodies in the rocks have lost more or less of their vegetable or animal matter, and, in many cases, are entirely changed to stone, they are often called Petrifactions—a term frequently used as synonymous with fossils; the latter term is preferable, however, because it includes all remains dug from the rocks, whether they have undergone complete petrifaction or not.

Precisely how petrifaction goes on, cannot well be explained. It consists in the substitution of mineral for animal or vegetable matter. But so perfectly are the form and structure of the organisms preserved, it is probable there is a constant interchange of particles between the organism and the adjacent mineral substance. That is, as fast as a particle of the organic body disappears, a particle of mineral matter takes its place. Whatever the process may be, the fact is established that petrifactions have taken place extensively in past times, and that they take place at the present day. The organic body is converted into lime, silica, pyrites, or other substance, according to the material in which the organism is imbedded, or which is disseminated in the surrounding rock.

FOSSILS.

Some fossils, though buried for ages, have not lost all their animal matter, as can be shown by chemical analysis.

The amount of organisms in the rocks is truly astonishing. In many parts of the state of New York, and throughout a large part of the Great Basin of the Mississippi, the rocks are filled with corals, crinoids, shells, and the remains of other marine animals. In fact, there are but few parts of the United States—and the same is true of other countries—where these fossils may not be found. In thousands of places in New York, and the states farther west and south, the great bulk of the rocks is composed of animal remains.

Fig. 101.

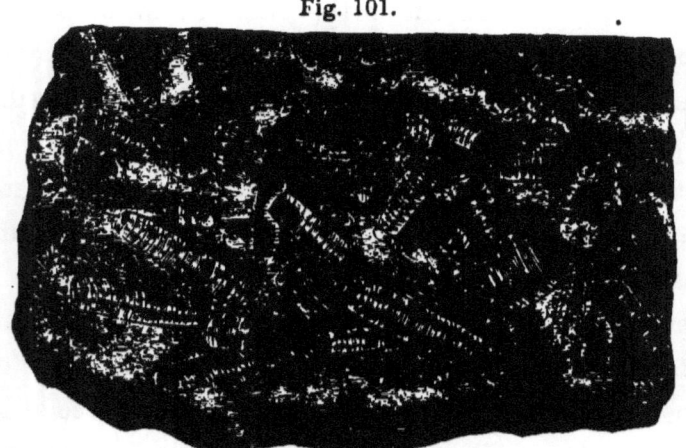

Perfect copy of Limestone, filled with Stems of Crinoids, near Lockport, N. Y.

In these regions, every blow of the geologist's hammer is sure to reveal a shell, a stem of a crinoid, a branch or mass of coral, or perhaps a crustacean, so strange in form that the observer at once refers it to an age long since gone by.

In many countries, and in some parts of our own—as at Sunderland, Mass.—the rocks are filled with fishes, so perfect in outline that they can readily be referred to their true place in the zoölogical scale. The rocks of Great Britain, and those of the continent of Europe, contain bones of gigantic reptiles, and bones of birds; and every country on the globe has its fossil mammals.

Beds of great thickness are not unfrequently found composed wholly of the remains of microscopic organisms. Richmond, Va., stands upon a bed of this kind, which is fifteen or twenty feet thick. The well-known polishing slate of Bilin, Germany, is composed of the shields of organisms so small that the remains of forty-one thousand millions are contained in a cubic inch of the stone.

Fossils occur in the rocks of the deepest valleys, and in those of the highest mountains. A large part of the Jura Mountains is composed of coral; and fossils occur in the Alps at the height of 8000 feet, and in the Andes and Himalayas at the height of 16,000. As a matter of fact, they abound in the rocks to the depth of eight or ten miles.

We may safely say, that, in many districts where fossils occur, there are more individuals, and sometimes more species in the rocks than living species upon the surface of the same territory; and all, except the more recent fossils, represent plants and animals which are specifically different from those of the present day.

Although fossils, as the Trilobites of Quincy and Braintree, Mass., and various organic remains found at the White Mountains, N. H., occur in rocks more or less metamorphic, they are not found in true granite, or other unstratified formations.

Dendrite, delicate expansions of mineral matter, closely resembling plants, is often found on breaking open all kinds of rocks, stratified as well as unstratified. These imitations of vegetable forms result from the infiltration of mineral matter into minute fissures, and must not be confounded with organic remains.

Representatives of all the great Branches and Classes of the Vegetable, and of the Animal Kingdom, have been found in the rocks. Thirty thousand or more species have been noticed, and thousands have been carefully examined, and their characters minutely recorded. But the work has only begun.

At the death of Cuvier, less than one hundred species of fossil fishes had been described; through the labors of Agassiz, mainly, the number soon reached 1600 or 1700. Similar progress has been made in the examination of other classes of the Animal Kingdom; and the rapid discovery of new species leaves little room to doubt that the fossil species of animals equal the living.

What an interesting fact,—that Nature has embalmed her subjects, and handed them down to us so perfectly preserved that we are able to get a glimpse, at least, of the phases of life during all the past ages of the world!

These fossils show that all the rocks in which they occur were once in a soft state, like the sand and mud at the bottom of our present waters; and occurring in successive layers, they teach us that each layer once constituted the upper surface, no matter to what depth it may now be below it. They show that the highest mountains have once been the ocean's bottom, and that, too, for a long time, since their sides are filled with corals, crinoids, shells, and other organisms, that could have grown only in the sea.

Animals and plants of the present time have well-marked characteristics, according as they are terrestrial, fresh-water, or marine. So it was in past times, and hence a careful study of fossils gives much information in regard to the early physical geography of our planet. We may learn what parts were under the ocean at a given time, what parts were estuaries, and what fresh water.

SECTION II.

CLASSIFICATION OF THE ROCK FORMATIONS.

IN a previous chapter it was stated that rocks are of two classes, Stratified and Unstratified—a classification based on the structure of the rocks. Considered in respect to fossils, rocks may also be divided into two classes—Fossiliferous, and Non-Fossiliferous. This classification corresponds somewhat, but not wholly, to the last.

The FOSSILIFEROUS comprise all the rocks that contain fossils—that is, all the stratified rocks, except gneiss, mica slate, hornblende, slate, &c. The NON-FOSSILIFEROUS comprise all the unstratified rocks, and such of the stratified as were formed before life was introduced upon our planet.

All the rocks formed while the conditions of the earth were essentially the same, constitute one great System. Every system is a record of the age in which that system or group of rocks was formed, and all the rock systems studied in their chronological order, reveal a history of our globe from earliest time. So true it is that Nature is her own historian.

The systems of fossiliferous strata, very generally acknowledged, named from the newest to the oldest inclusive, are the following: Alluvium and Drift, Tertiary, Cretaceous, Oölitic or Jurassic, New Red Sandstone, Carboniferous, Old Red Sandstone or Devonian, and Silurian. These are subdivided into many formations.

Each of the above systems is characterized by peculiar fossils, the same species being rarely common to any two of them; but the same orders, and sometimes the same genera, are found in two contiguous systems.

These groups of rocks, then, with their imbedded fossils, represent the great Life Periods of the globe. One race of plants and animals has occupied the surface of the earth, for long ages, and then passed away; and another race, different and higher in rank than the one before it, has taken its place;—and this has been repeated as many times, at least, as there are groups indicated above, and probably many more.

Below the Fossiliferous Rocks are the Non-fossiliferous, embracing the oldest slates, gneiss, granites, &c., as specified above.

The following table shows the Classification of the Rock Formations, as adopted in this treatise, and also the divisions and groups which have been made of the same formations, by some other writers.

CLASSIFICATION OF THE ROCK FORMATIONS.

TABULAR VIEW OF THE CLASSIFICATION OF BOTH STRATIFIED AND UNSTRATIFIED ROCKS.

FOSSILIFEROUS ROCKS.	*Modern or Quaternary Rocks.*	Alluvium and Drift.	Post-Pliocene.
	Tertiary Rocks.	Tertiary.	Pliocene. Miocene. Eocene.
	Secondary Rocks.	Cretaceous.	Chalk. Greensand. Wealden.
		Oölitic or Jurassic.	Oölite. Lias.
		New Red Sandstone.	Trias. Permian.
	Paleozoic Rocks.	Carboniferous.	Coal Formation. Conglomerate. Mountain Limestone.
		Old Red Sandstone, or Devonian.	Upper Old Red. Lower " "
		Silurian.	Upper Silurian. Lower "
NON-FOSSILIFEROUS ROCKS.	*Azoic Rocks.*	Oldest Metamorphic, and First-formed Rocks.	Quartz Rock, Mica Slate, Hornblende Slate, Gneiss, and the oldest Granites.

142 CLASSIFICATION OF THE ROCK FORMATIONS.

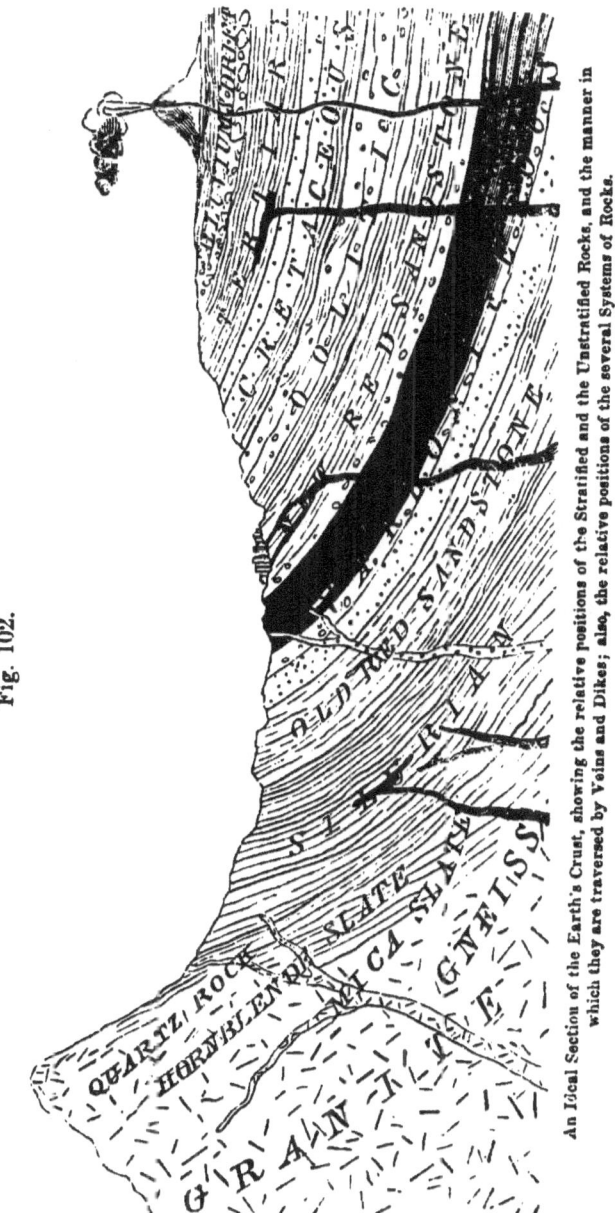

Fig. 102.

An Ideal Section of the Earth's Crust, showing the relative positions of the Stratified and the Unstratified Rocks, and the manner in which they are traversed by Veins and Dikes; also, the relative positions of the several Systems of Rocks.

CHAPTER VIII.

BRIEF DESCRIPTION OF THE SEVERAL SYSTEMS OF FOSSILIFEROUS ROCKS.

SECTION I.

SILURIAN SYSTEM.

The Silurian is the lowest system of rocks in which fossils are positively known to occur. Its name is derived from the ancient Roman designation of the part of England, where it was first scientifically observed and studied. It consists of two great divisions, the Upper and the Lower Silurian; but it will be sufficient for our purpose to consider both together. For the most part our remarks will be of a general nature, applicable to the whole system, modified, however, by such specific statements as the case demands.

The Silurian is represented in every country on the globe. In North America, it abounds in the Hudson's Bay basin, in the north-eastern parts of New England, in the valley of the St. Lawrence, in the central, south-eastern, and western portions of New York, and thence southward along the Alleghany Mountains. Rocks of this system appear in Ohio, at Cincinnati and its vicinity, and at Nashville, Tennessee. The Silurian is also extensively represented in England, Norway and Sweden, Belgium, Germany, Russia, and Australia.

Like every other system, the Silurian embraces a great variety of rocks. The student must not expect always to find the rocks in any system agreeing in mineralogical composition, throughout considerable depths, and over wide areas.

The prevailing rocks of the Silurian are sandstones, slates, conglomerates, and limestones.

The rocks of this system are not only of great geographical extent, but also of immense thickness; the aggregate of the several stages being not less than 30,000 feet.

The Silurian system abounds with fossils. Besides plants belonging to the Algæ tribe, more than one thousand species of animals have been discovered, and the number of individuals is immense.

This system contains representatives of the four great Branches of the Animal Kingdom—Radiates, Molluscs, Articulates, and Vertebrates.

RADIATES are represented by two Classes— Polyps and Echinoderms—and by the lowest orders of these classes.

SILURIAN SYSTEM.

POLYPS are represented of course only by their hard parts, to which the name Coral is applied. In many stages of the Silurian, this occurs in the greatest abundance, and in a great number of species. More species of coral occur in the rocks of this system in the state of New York, than are now living on the coast of Florida.

Fig. 103.

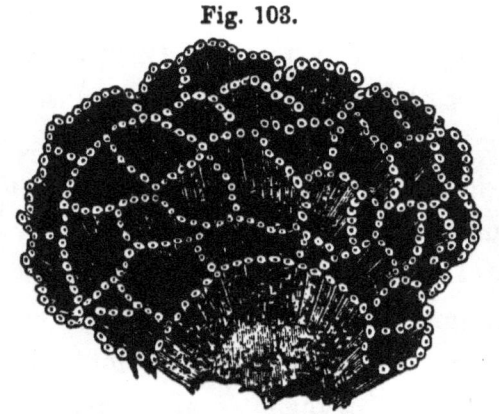

Silurian Coral.—*Catenipora escharoides.*

ECHINODERMS are abundant, but the representatives

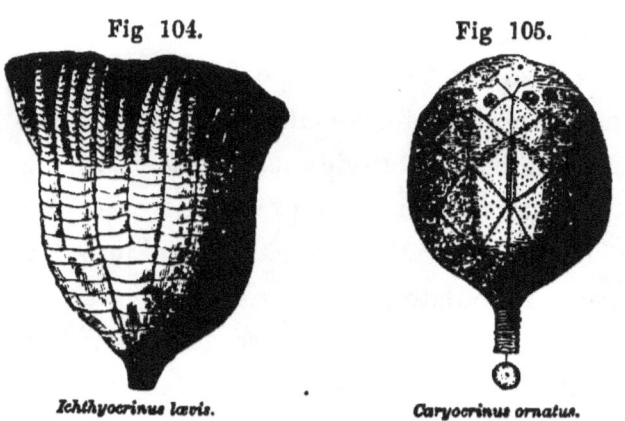

Fig 104.   Fig 105.

Ichthyocrinus lævis.   *Caryocrinus ornatus.*

Some of the forms of Silurian Crinoids.

146 SILURIAN SYSTEM.

of this class belong almost wholly to the Order of *Crinoids*.

MOLLUSCS appear in abundance in all their Classes—Acephals, Gasteropods, and Cephalopods.

The ACEPHALS are abundant in the form of *Brachio-*

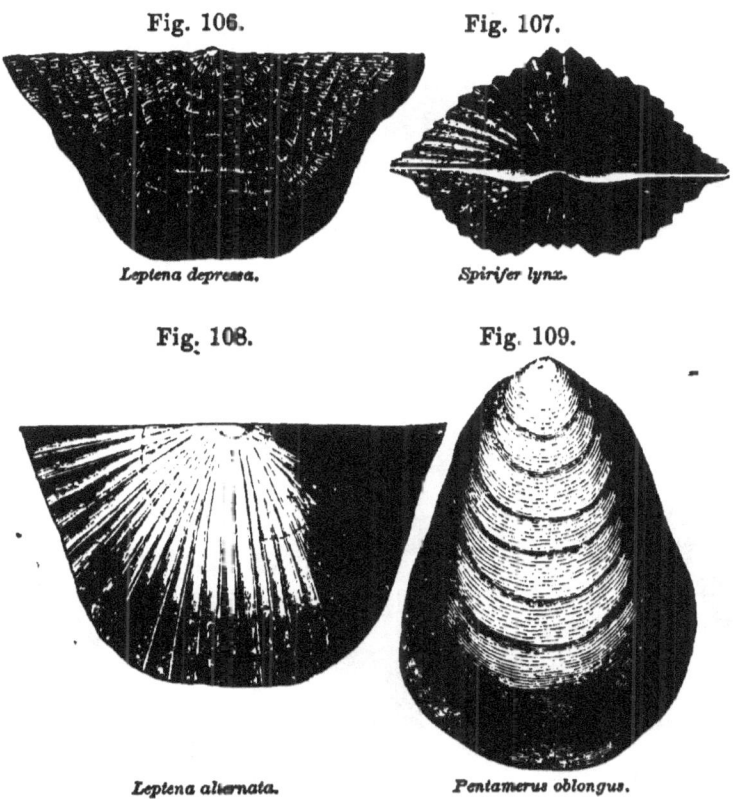

Fig. 106. *Leptena depressa.*
Fig. 107. *Spirifer lynx.*
Fig. 108. *Leptena alternata.*
Fig. 109. *Pentamerus oblongus.*

Some of the common Silurian Acephals, or Bivalves.

pods, the only fossils yet found in the Potsdam sandstone, the oldest member of the Silurian in this country.

GASTEROPODS are represented by many species.

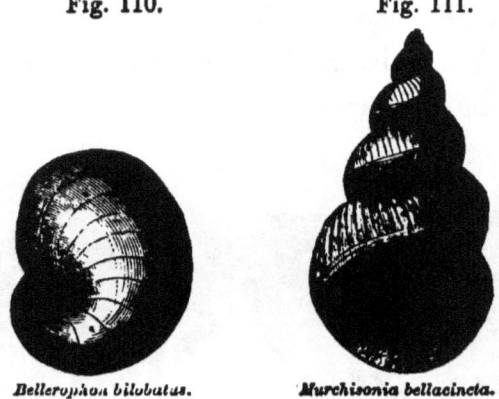

Fig. 110.   Fig. 111.

Bellerophon bilobatus.   *Murchisonia bellacincta.*

Some of the common forms of Silurian Gasteropods.

CEPHALOPODS are represented by both straight and coiled chambered shells. The straight shells are

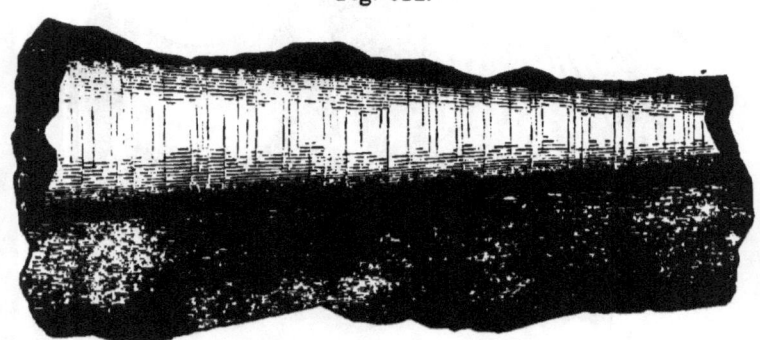

Fig. 112.

Portion of an Orthoceratite imbedded in Limestone.

Silurian Cephalopod.

called Orthoceratites, and are so abundant in some parts of the system as to touch and overlie each other rendering it almost impossible to secure a perfect specimen. They occur of all sizes, from a few inches in length to ten feet long, and a foot in diameter, as in the Black River Limestone, N. Y. The Cephalopods of this, and the three succeeding periods, are all of the Order of *Tetrabranchiates*.

ARTICULATES appear mainly in the form of Crustaceans, which occur in great numbers. Because none or but very few worms are found, we must not infer that they did not exist; for the softness of their bodies would render them less likely to be preserved than animals with harder parts.

CRUSTACEANS are represented by a very interesting

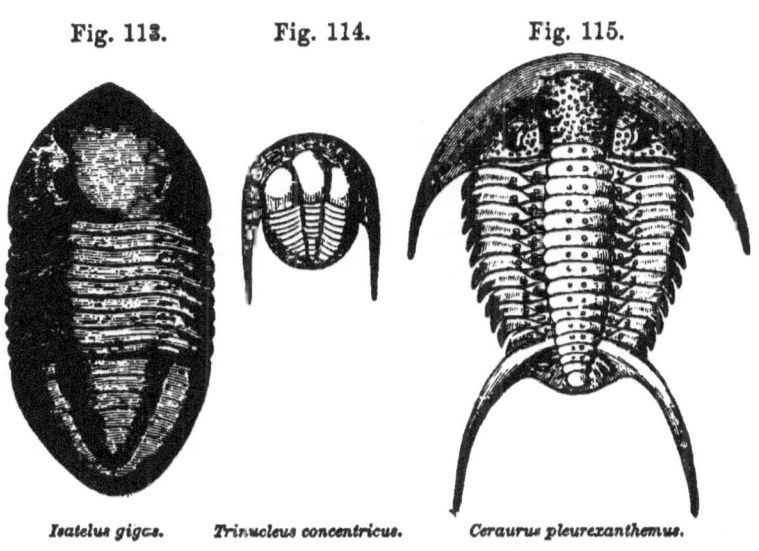

Fig. 113.   Fig. 114.   Fig. 115.

Isatelus gigas.   *Trinucleus concentricus.*   *Ceraurus pleurexanthemus.*
Some of the forms of Silurian Trilobites.

race of animals called *Trilobites*. Their remains occur

in the greatest abundance, and they have the widest possible geographical distribution. In many cases the same species are found in North America, Europe, and Australia.

These ancient crustaceans remind us of the Limulus, or Horse-shoe Crab, to which they are more nearly allied than to any other living species, excepting, perhaps, a group of crustaceans known to naturalists under the name of Phyllopoda. Trilobites had the power of rolling themselves into a ball, and many of them are found in that position.

In the Trenton Limestone, at Trenton Falls, N. Y., they occur in the greatest profusion, mixed with shells, crinoids, and corals. In some cases the eyes of trilobites are preserved, and this fact shows that there was light during the Silurian period, and that it sustained the same relation to animal life then as now.

VERTEBRATES are represented in the Silurian only by FISHES. No batrachians, reptiles, birds, or mammals have left their remains in this system of rocks; which is conclusive evidence that they had not as yet appeared upon the earth. Nor have we yet authentic evidence that fishes occur in the lower portions of the Silurian; but it is well established that their remains occur in some of the higher beds. So the proof is positive, that the four great Branches, or Types, of the Animal Kingdom, began their existence in the same great geological period, if not simultaneously, which is highly probable.

The fishes of this period are widely different from those of the present day. They were of the lowest order, having cartilaginous skeletons, and other marks of an inferior rank.

SECTION II.

OLD RED SANDSTONE, OR DEVONIAN SYSTEM.

This is the next great system, in the ascending order, above the Silurian, and, like it, of immense thickness. The term "Old Red Sandstone," was first given to the strata of this era, from the red color of the prevailing rocks where it was first studied. Later, "Devonian" was applied, from Devonshire, where this system is rich in fossils. Both terms are now used indiscriminately.

In this country, the Devonian has its greatest development in New York and Pennsylvania, and thence, on a less extensive scale, it extends south and west.

In Europe, it is largely represented in Russia, in Western Europe, and especially in England and Scotland, being 10,000 feet thick in the latter country.

The rocks of this system are of various kinds;

they are mainly, however, dark red sandstones, and conglomerates, in the upper part, and slates, sandstones, and limestones, in the lower. In the state of New York the Old Red Sandstone is 14,000 feet thick, but is much thinner in the states farther west and south.

The fossils which fill some of the stages of this system, show that the ancient oceans in which these rocks were formed, swarmed with life. Many of the organic forms of this period bear a great resemblance to those of the Silurian, but differ specifically from them.

The number of fossils already noticed in the Devonian is about one thousand species. Plants are numerous, but not well preserved. Hugh Miller has detected Gymnosperms in the rocks of this era.

All the four Branches of the Animal Kingdom are represented.

RADIATES, as in the Silurian, appear in the form

Fig. 116. Fig. 117.

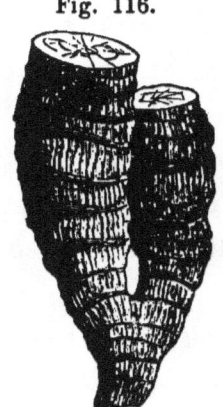
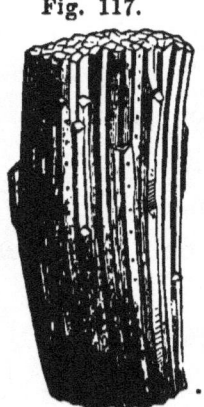

Cyathophyllum dianthus.   *Favosites Gothlandica.*

Devonian Corals.

of *Corals* and *Crinoids*. Some of the beautiful specimens of the former are represented by figs. 116 and 117.

MOLLUSCS are found in all their Classes, and present many unique forms.

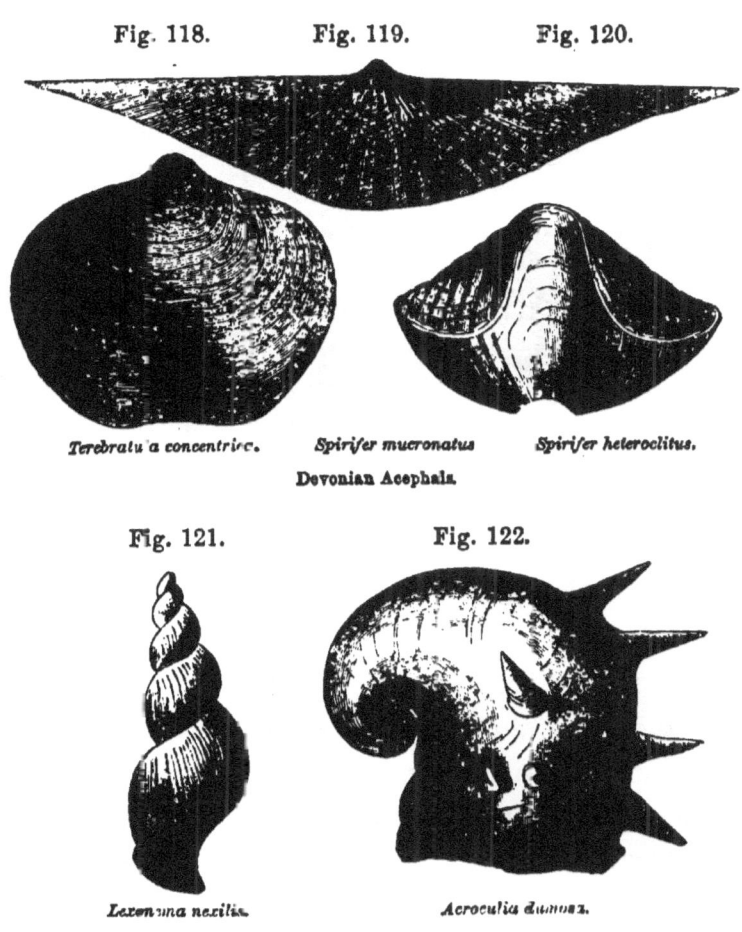

Fig. 118.   Fig. 119.   Fig. 120.

Terebratula concentrica.   *Spirifer mucronatus*   *Spirifer heteroclitus.*

Devonian Acephala.

Fig. 121.   Fig. 122.

Loxonema nexilis.   *Acroculia dumosa.*

Devonian Gasteropoda.

ARTICULATES are represented by WORMS and

CRUSTACEANS. *Trilobites,* as in the last system, are the only representatives of the latter, and they are less abundant than in the previous period. Fig. 123 shows a species of Trilobite found in all places where the Devonian occurs.

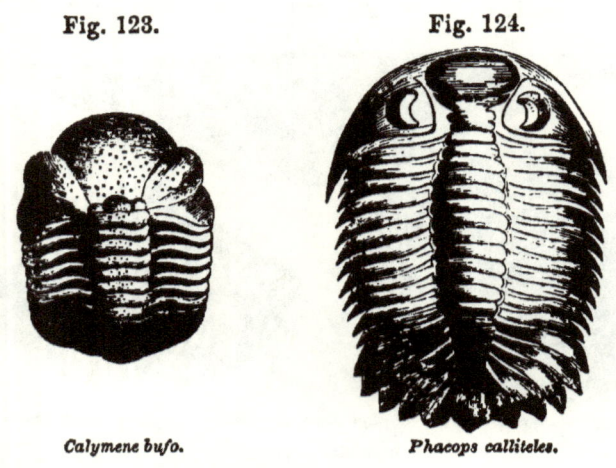

Fig. 123.   Fig. 124.

Calymene bufo.   *Phacops calliteles.*

Devonian Trilobites.

VERTEBRATES are represented by Fishes and by Batrachians; though but few species of the latter have been found. FISHES abound in this system. Introduced by the single order of Placoids, during the Silurian, they appear in this period in the additional order of Ganoids; and the two expand so that, notwithstanding the numerous corals, bivalves, and trilobites, Fishes constitute the leading feature in the Devonian fauna.

These two orders—the Placoids and the Ganoids—

not only comprise all the Fishes of this and the preceding period, but also those of the Carboniferous, the

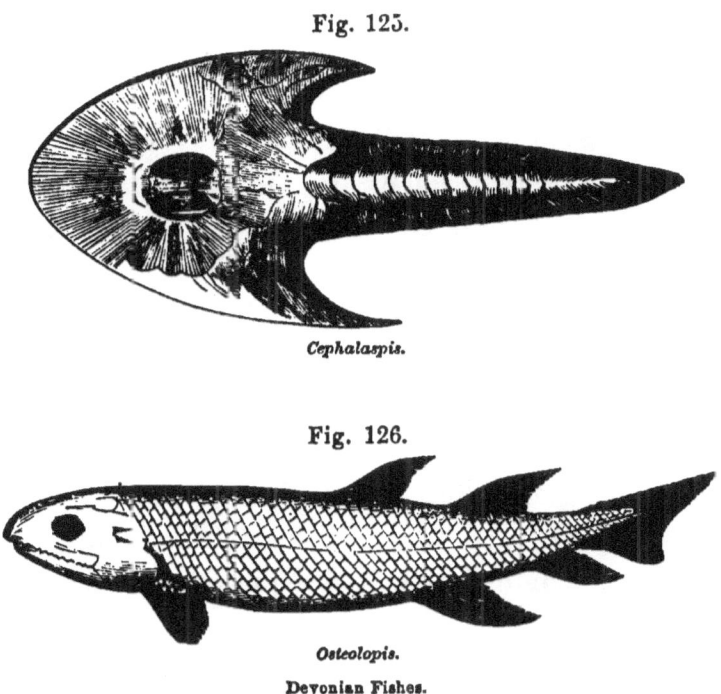

Fig. 125.

Cephalaspis.

Fig. 126.

Osteolopis.
Devonian Fishes.

New Red Sandstone, and the Oölitic. Not till the Cretaceous do the two highest orders of this Class make their appearance upon earth.

SECTION III.

CARBONIFEROUS SYSTEM.

This system is next in the ascending order above the Old Red Sandstone. Its geographical area is very wide, being found in **almost** every country on the globe. It is very extensively represented in the United States, and in the British Provinces of North America. In Europe it is largely developed in Great Britain and on the Continent.

The rocks of this system are limestones, sandstones, conglomerates, and shales, among which, in many regions, are seams or beds of mineral coal, from a fraction of an inch to 40 or 50 feet in thickness. The lower part of this system is occupied by an extensive formation, called the Carboniferous or Mountain Limestone. Above the Mountain Limestone, we find the true Coal Formation, in most places resting on Conglomerate. The Mammoth Cave, and many other great caves of the Western States, occur in Mountain Limestone. This rock also forms the high bluffs along many of our Western rivers.

Both fossil plants and animals are abundant in the Carboniferous. The plants, however, are mostly confined to the Coal Formation, while the animal remains are most abundant in the Mountain Limestone.

156 CARBONIFEROUS SYSTEM.

Fig. 127.

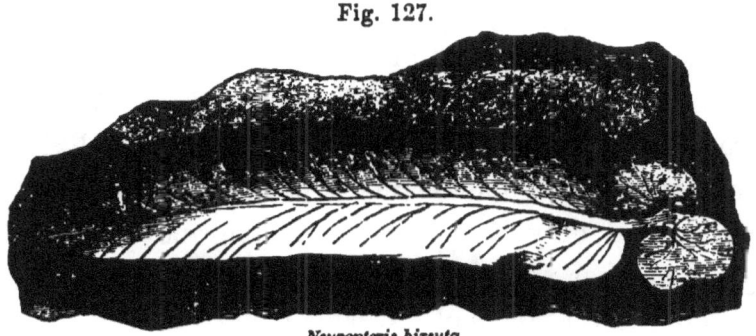

Neuropteris hirsuta.

A Fern of the Coal Period.

Fig. 128.

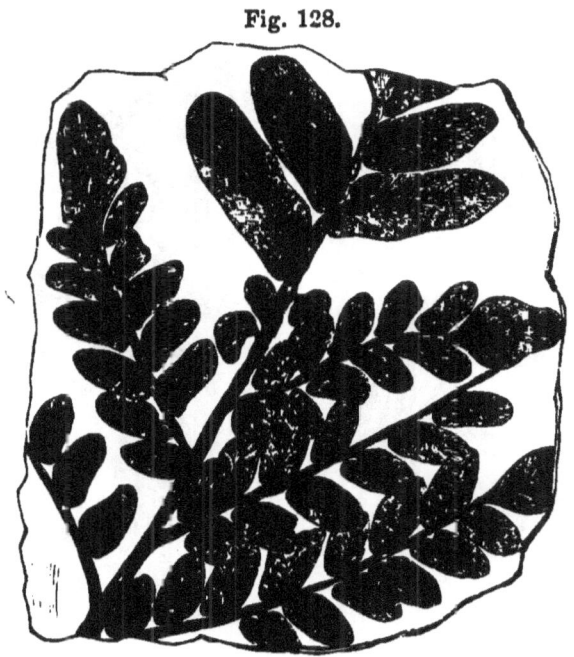

Neuropteris Loshii.

A Fern of the Coal Period.

The Carboniferous was the great plant period of ancient times. The flora of this era was composed mainly of Cryptogamous plants; although Conifers, or those allied to the Coniferæ, were also abundant. From 250 to 300 species of *Ferns* have been obtained from the Coal Formation. There are only 50 living species of Ferns indigenous to the Northern United States, and only 60 species in all Europe. Yet the great number specified above, has already been discovered in the rocks of the same countries.

Fig. 129.

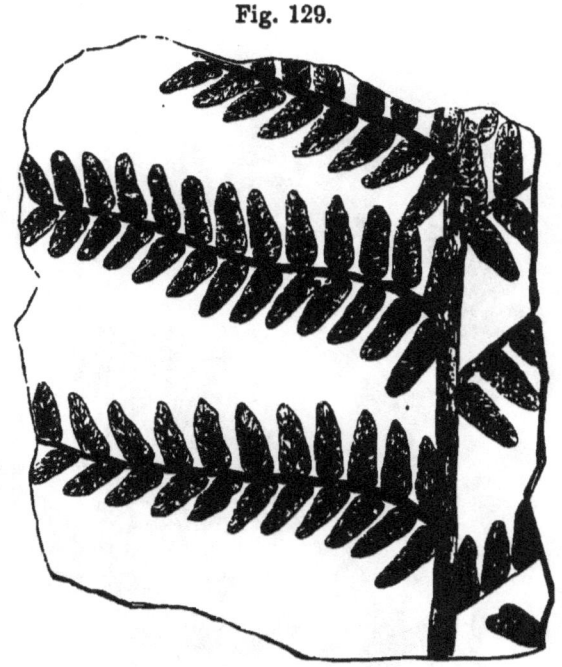

Pecopteris distans.

A Fern of the Coal Period.

Asterophyllites, a family of plants of doubtful affinity, are common in the coal formation. Figs. 130

and 131 show two of the common species of this family.

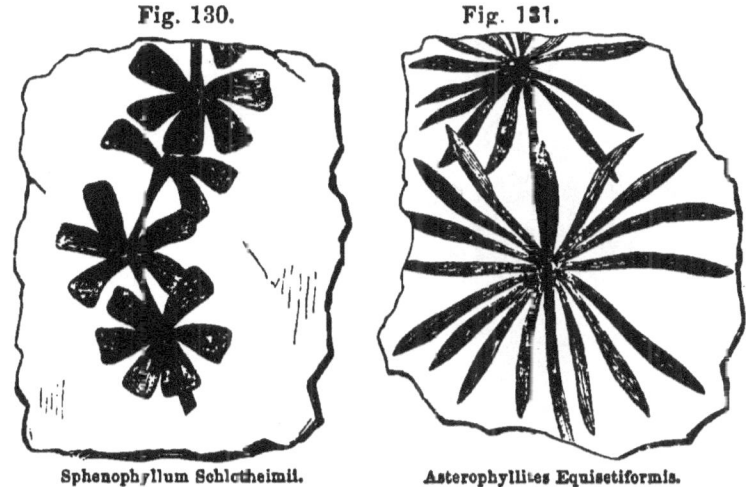

Fig. 130. Sphenophyllum Schlctheimii.

Fig. 131. Asterophyllites Equisetiformis.

Calamites resembling gigantic Equisetaceæ, but also of doubtful affinity, are abundant.

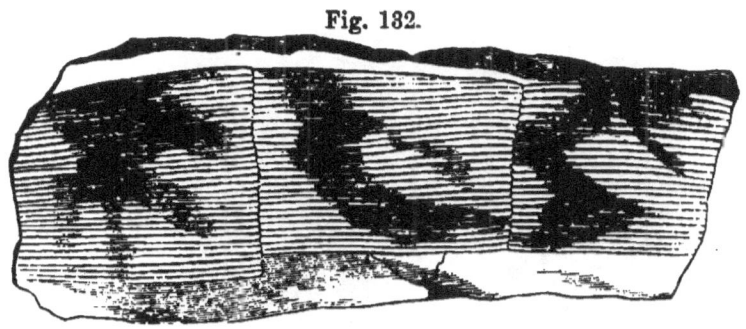

Fig. 132. Stem of Calamites imbedded in shale.

Lepidodendra, gigantic Club-Mosses, or closely allied to them, are common in the rocks of the coal period. The bark of two species is represented by Figs. 133

CARBONIFEROUS SYSTEM.

Fig. 133.

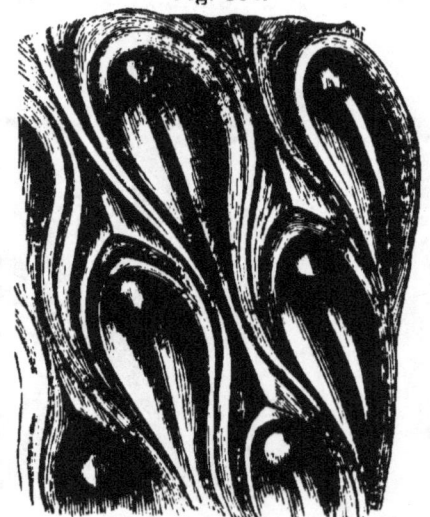

Lepidodendron obovatum.

Fig. 134.

Lepidodendron obtusum.

and 134. It is common to find Lepidodendra 20 or 30 feet high, and the remains of one in the rocks at Carbondale, Pa., show that it was at least 75 feet high, and two feet in diameter.

The living Club Mosses, in the temperate zones, are trailing plants, or rise to the height of only a few inches; and even in the torrid zone, the largest are not above three feet high.

Trees, called *Sigillaria* from their peculiarly marked stems, are very abundant in the rocks of this era. A portion of the stem of one of the most common species is represented by the following cut.

Fig. 135.

Sigillaria reniformis.

They grew from 30 to 70 feet high. The woody portion, in many cases, has decayed, leaving the bark, which is often found pressed together, forming two thin layers of pure coal. Wherever the trees remain upright, that is, at right angles to the planes of stratification, as they do in great numbers in the coal fields of Nova Scotia, the trunk preserves the cylindrical form.

Between thirty and forty species of Sigillaria have already been discovered, and they seem to be the most numerous of all the trees of the coal period. It is difficult to refer them to their true place in the botanical scale. Though having some characters which connect them with ferns, they are probably as nearly related to Cycads as to any living family.

The root of these trees was for a long time described under the name of Stigmaria, from the impression that it was a distinct plant. But in many instances, within a few years, the root and stem have been found in contact, thus proving Sigillaria and Stigmaria parts of one plant.

Fifteen hundred or more species of animals have been found in the Carboniferous system. The Mountain Limestone is especially rich in fossils, representing the Animal Kingdom in all its branches. Many species unknown in the previous systems are abundant in this.

RADIATES appear in numerous species of *Corals* and *Crinoids*.

Both are very abundant in the Mountain Limestone of the Western States. Some of the beautiful specimens of crinoids are represented by Figs. 136, 137, 138.

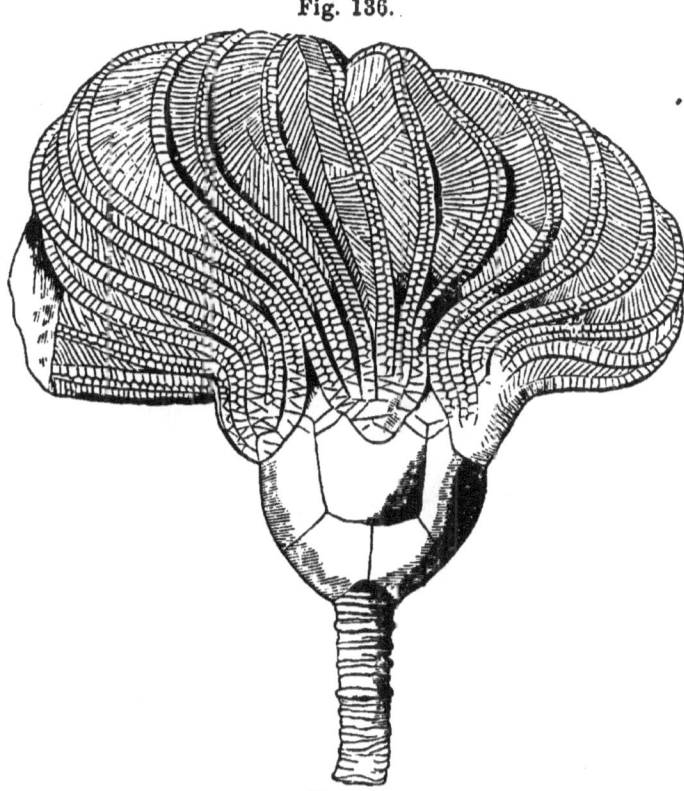

Fig. 136.

Platycrinus.

Fig. 137.

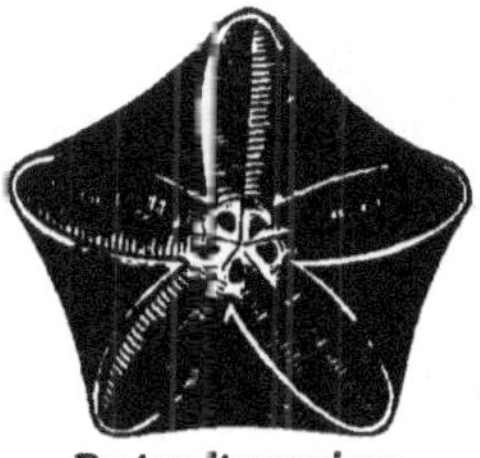

Pentremites cervinus.

Fig. 138.

Pentremites pyriformis.

Carboniferous Crinoids.

Corals occur in great perfection. A characteristic species is represented by Fig. 139.

Fig. 139.

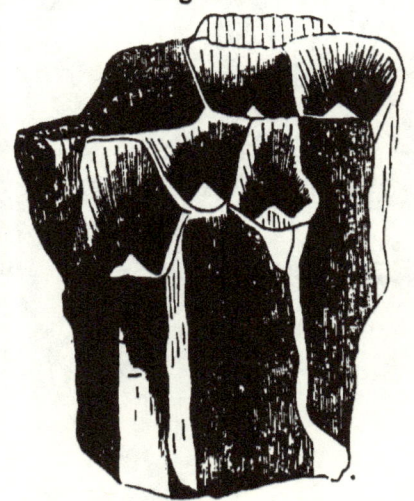

A Carboniferous Coral.—*Lithostrotion basaltiforme.*

MOLLUSCS are represented in all their Classes. The family called Productidæ reaches its highest expression in this era.

Fig. 140.

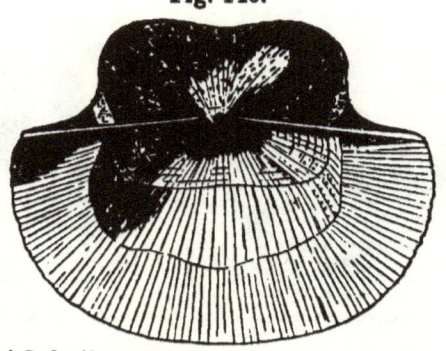

A Carboniferous Mollusc.—*Productus semireticulatus.*

CEPHALOPODS, represented by both straight and coiled chambered shells, are abundant in the Mountain Limestone.

ARTICULATES are represented by Trilobites and INSECTS; the latter being, so far as we know, the first land inhabitants of our earth. Trilobites, so abundant in the Silurian and Devonian, appear here in only a few species; and above this system not one is found. The race of Trilobites ended with the Carboniferous period.

VERTEBRATES appear in the form of Fishes and Batrachians.

FISHES, as in the rocks of the previous period, are exceedingly abundant.

Fig. 141.

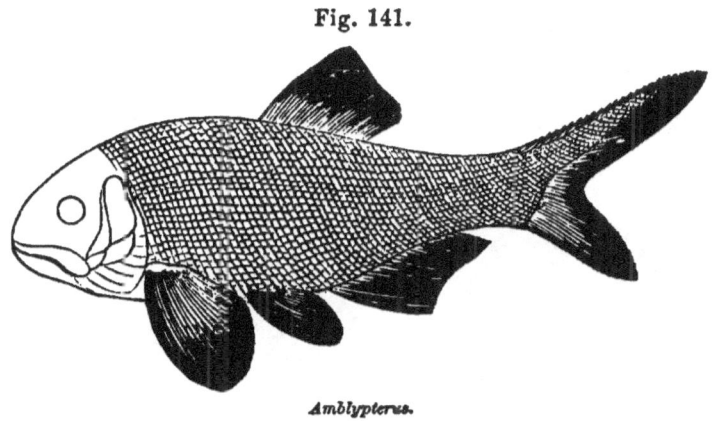

Amblypterus.

Carboniferous Fish.

BATRACHIANS have left their foot-prints on the carboniferous rocks in Nova Scotia, New Brunswick, and Pennsylvania.

CARBONIFEROUS SYSTEM.

MORE SPECIFIC REMARKS ON THE COAL FORMATION.

In most regions where the coal formation occurs, there are several beds or seams of coal with layers of shales, sandstones, conglomerates, and sometimes limestone, intervening. In some cases, the coal-beds are separated by only a few feet; in others, several hundred feet of rocks intervene. In the district called the Joggins, in Nova Scotia, the coal formation is 14,000 feet thick, and contains seventy-six beds or seams of coal; only part of these, however, are of workable thickness. In the basin of the Schuylkill, Pennsylvania, there are about fifty seams, twenty-five of which exceed three feet each in thickness. One bed, at Mauch Chunk, is from forty to fifty feet thick.

The way in which coal occurs interstratified with conglomerates, sandstones, and shales, may be readily

Fig. 142.

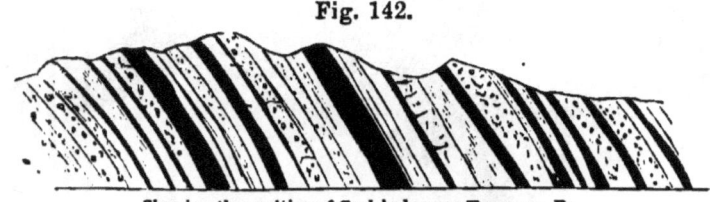

Showing the position of Coal-beds near Tamaqua, Pa.

understood from the accompanying cut, where the black layers represent the coal-beds.

There are two principal kinds of mineral coal—Anthracite and Bituminous.

Anthracite is mostly without bitumen, very hard, with a high lustre, often iridescent, and burns with a pale blue flame.

Bituminous coal abounds with bitumen, is softer than anthracite, with little lustre, and burns with a bright flame. It appears in many varieties, one of which is the well-known cannel coal.

There is every possible grade between the true anthracite and that which may be properly called bituminous coal.

Mineral coal has been found in nearly all parts of the world, but more has been discovered in North America than in all other countries.

Great Britain has 12,000 square miles of coal-field, and the continent of Europe about 10,000 square miles; while the area of coal-fields in Nova Scotia, New Brunswick, and vicinity, is 8000 square miles, and the area of those in the United States more than 200,000.

The eastern half of North America contains five great coal-fields; that of Nova Scotia, New Brunswick, &c., embracing, as stated above, 8000 square miles; the Great Appalachian coal field, extending from Ohio and northern Pennsylvania to Alabama, embracing 80,000 square miles; the Michigan coal-field, 15,000 square miles; the Indiana, Illinois, and Kentucky coal-field, 50,000 square miles; and the Iowa and Missouri coal-field about 60,000 square miles.

CARBONIFEROUS SYSTEM.

Professor H. D. Rogers, in his recent able Report on the Geology of Pennsylvania, states the approximate amount of coal in each of the great coal-fields of the world. I give his figures below.

	Tons.
Belgium,	36,000,000,000
France,	59,000,000,000
British Isles,	190,000,000,000
Pennsylvania,	316,400,000,000
Great Appalachian Coal-Field,	1,387,500,000,000
Ind., Ill., and Ky., " "	1,277,500,000,000
Ia., Mo., and Ark., " "	739,000,000,000
All the coal-fields of North America,	4,000,000,000,000

It will be observed that the amount of coal in North America is twenty-one times as much as that of Great Britain.

It is well established that the coal-beds are of vegetable origin. This might be inferred from the fact that coal is mainly Carbon, which substance forms from one-fourth to one-half of all the vegetation on the globe. But, prepared in very thin slices, coal shows its vegetable structure under the microscope, and often even to the naked eye.

The vegetation which formed the coal-beds, accumulated, by the slow growth of plants, just as peat is now forming in peat-bogs. Our peat-bogs are only incipient coal-beds, and need only to be submerged, and covered with great depths of strata, in order to become genuine coal. The varieties of coal offer no objection to this statement; all being alike in their essential element, carbon, and differing only as they have been subject to different influences. Anthracite is only bituminous coal, which, by heat and pressure, has been deprived of its volatile matter; and the varieties between the true bituminous coal and the true anthracite, have resulted from different degrees of heat and pressure to which the vegetation that formed them has been subjected. Anthracite is found only in situations where the strata have been heated and disturbed since the coal was formed, and the coal becomes more and more bituminous as we recede from such regions. The coal in Massachusetts and Rhode Island, and in the eastern part of Pennsylvania, is anthracite; but in the Great Mississippi Basin, where the strata are little disturbed, only bituminous varieties are found.

The succession of coal-beds, with beds of rock intervening, which is found in most coal-fields, implies that the lands have undergone successive elevations and depressions, during vast periods of time. Such vertical movements, we shall learn in a subsequent chapter, are still going on in various parts of the world.

Iron, though common in other rocks, abounds in the Carboniferous System, and frequently occurs in close proximity to the coal. This is the case in England. In the United States there is iron enough to supply the world. Coal, iron, and limestone are all abundant in the same geological system; and in smelting iron both the others are required. Such sources of national prosperity cannot be the result of accident. Are they not proofs of Design and Divine Benevolence?

SECTION IV.

NEW RED SANDSTONE SYSTEM.

This system is next, in the ascending order, above the Carboniferous. It consists of two formations or stages, the Upper, and the Lower New Red. The former is often called the Trias, from the three divisions which it exhibits in Germany, where it was first studied; and the latter, the Permian, from the province of Perm, where this stage is largely developed.

This system is composed of red sandstones and shales in the upper part, and magnesian limestones in the lower part.

The sandstones and shales of Connecticut River

valley, New Jersey, Virginia, and North Carolina, and the Permian limestones and shales of Kansas, represent the New Red in this country.

This system is well represented in Europe; and both in Great Britain and on the Continent, its upper portions contain extensive deposits of gypsum and rock-salt.

The New Red Sandstone is almost everywhere penetrated by, and interstratified with igneous rocks; and in many cases they overlie the strata in immense masses of greenstone and basalt.

In many places in the Connecticut River valley, the layers of rocks of this period are covered with ripple-marks, as distinct as those upon the shores and shallow estuaries of to-day.

Though plants and all the Branches of the animal kingdom are represented in the New Red Sandstone, the fossils which represent the VERTEBRATES of this period are of special interest.

Besides FISHES, which are abundant, evidences of other Vertebrates are found in the foot-prints which occur in the rocks of this era, both in this country and in Europe. But no other place on the globe is so much noted for fossil foot-prints as the valley of the Connecticut. Thanks to Professor Hitchcock, Dr. Deane, and others, we are pretty well acquainted with the paleontology of this beautiful valley.

From the northern part of Massachusetts to New

Haven, Ct., a distance of 90 miles, the sandstone of this valley reveals foot-prints in the greatest abundance. They occur in great numbers, on successive layers, to the depth of many feet. Professor Hitchcock, who, for nearly a quarter of a century, has studied, with untiring zeal, the Connecticut river sandstone, finds that it contains the tracks of more than 100 species of animals, of which number Batrachians, Reptiles, Birds, and Marsupialoids constitute the principal part.

Some of the tracks are very small, while others are of enormous dimensions. Slabs of sandstone, containing nearly all the species of

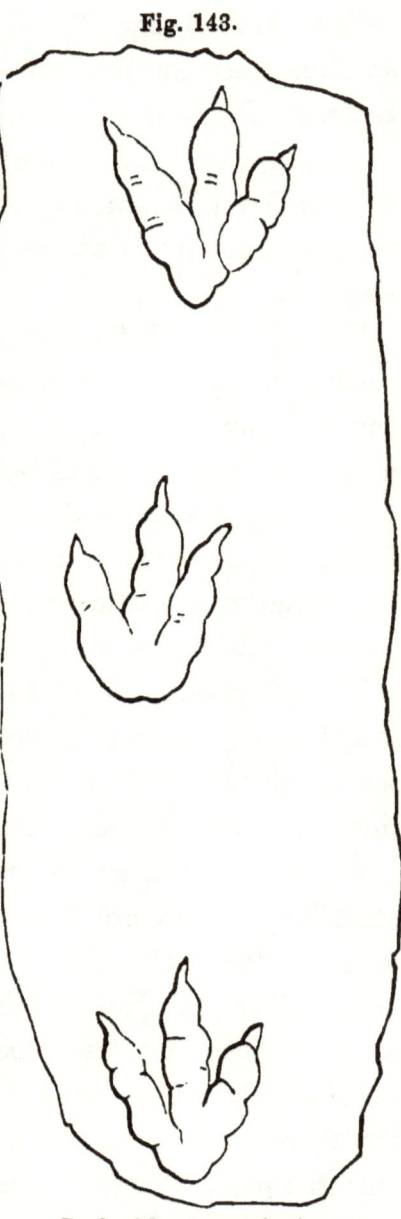

Fig. 143.

Tracks of Brontozoum giganteum.

Fig. 144.

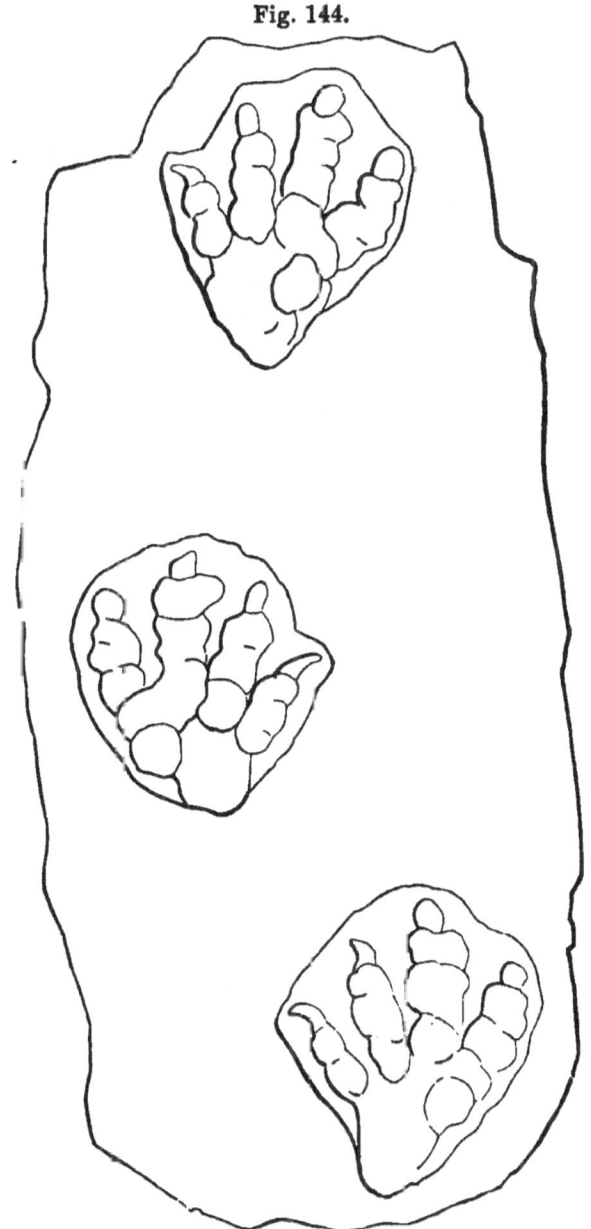

Tracks of Otozoum Moodii.

tracks, are deposited in the Ichnological Cabinet of Amherst College. Slabs from Northampton show foot-prints 18 inches long, and the stride from 3 to 5 feet! These, it is supposed, were made by a gigantic bird, which Professor Hitchcock has named *Brontozoum giganteum*.

Fig. 144 represents tracks found at South Hadley, and several other places in the valley of the Connecticut. These tracks are 20 inches long, and 15 inches wide! According to the distinguished geologist mentioned above, they were made by a huge Batrachian, which he has named *Otozoum Moodii*.

What a strange contrast between the present fauna of this valley, and that which held sway during the Red Sandstone period!

In many cases, the same foot-print extends through several layers; and thus a given specimen may be multiplied by splitting the layers apart. The upper surface shows the imprint, and the under surface the same track in relief. Professor Hitchcock has found many such specimens, and secured the layers by hinges so as to form a real "Stone Book,"—Nature's own Record of Sandstone days.

One of these specimens is represented by Fig. 145.

In numerous instances, impressions of rain-drops are found on the Connecticut river sandstone, and not unfrequently on the same slabs with the foot-marks. The meteorological phenomena of that distant age are thus registered with unerring certainty.

174 NEW RED SANDSTONE SYSTEM.

Fig. 145.

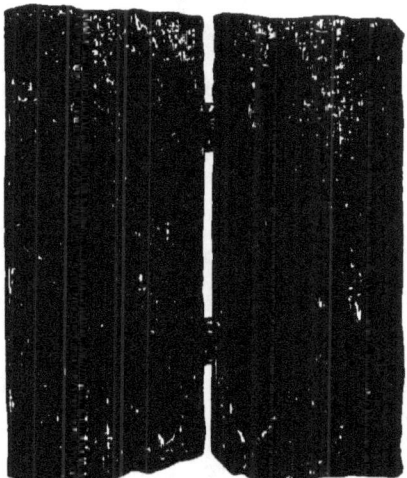

The same Tracks impressed, and in relief.

We learn from these tracks of the Connecticut valley, that the rocks upon which they occur, were formed in an estuary—along the shores and in the shallows of which the inhabitants of ancient times congregated.

In closing this section, I ought to say, that eminent geologists believe that the rocks of this country which are here described under the head of New Red Sandstone, belong to the Oölitic, or at least to the Lias, the lowest formation of the Oölitic System. But perhaps this point is not yet fully established.

SECTION V.

OÖLITIC SYSTEM.

This system, embracing the Lias, succeeds, in the ascending order, the New Red Sandstone. In this country it has been carefully studied only in Virginia, but it occurs also in the region of the Rocky Mountains. In England, it extends in a belt thirty miles broad, from Yorkshire to Dorsetshire. It is also extensively developed in the Jura Mountains, and hence is often called the Jurassic System.

The rocks of this period are limestones, clays, shales, and sandstones; and occasionally beds of coal. The well known lithographic stone of Solenhofen, Bavaria, belongs to the Oölitic.

The name Oölitic was given to the system from the egg-shaped nodules which abound in some of its limestones.

Plants were abundant in this era. They were mainly Ferns, Cycads, and true Conifers.

RADIATES are represented in all their Classes. Whole reefs of coral are found in the position where they grew. The formation in England called the Coral Rag, is mostly composed of coral, and hence its name.

ACALEPHS, or Jelly-fishes, are found for the first

time, in the rocks of this period. They occur at Solenhofen, Bavaria. This may or may not have been the time in geologic history, when they were introduced upon earth; for their perishable nature would render their preservation quite uncertain.

ECHINODERMS are abundant in the rocks of this era, and are represented by beautiful Crinoids, Star-fishes, and a great variety of Echinoids.

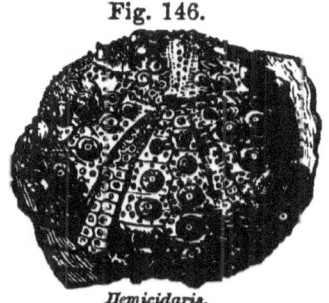

Fig. 146.

Hemicidaris.

Echinus of the Oolitic, with spines removed.

MOLLUSCS appear in all their Classes. The forms of the previous ages, however, have given place to a great number of new ones. Brachiopods, which have been very numerous in the preceding periods, are abundant here in only one family, the Terebratulæ.

CEPHALOPODS. It is in this period and the next, that this interesting class reaches its highest expression. The Orthoceras, so abundant in the earliest systems, have wholly died out; but Ammonites and Nautili, which occupied a comparatively subordinate place in the paleozoic fauna, appear here and in the next system, in the greatest abundance; and their highly varied and ornamented shells are not without great interest even to the casual observer.

The Ammonite differs from the Nautilus in having folded partitions, while those of the latter are simply curved, as seen in Fig. 81 on the 116th page.

OOLITIC SYSTEM. 177

Dibranchiates first make their appearance in this period. They are represented by a family called Belemnites, animals much resembling modern Cuttle-fishes.

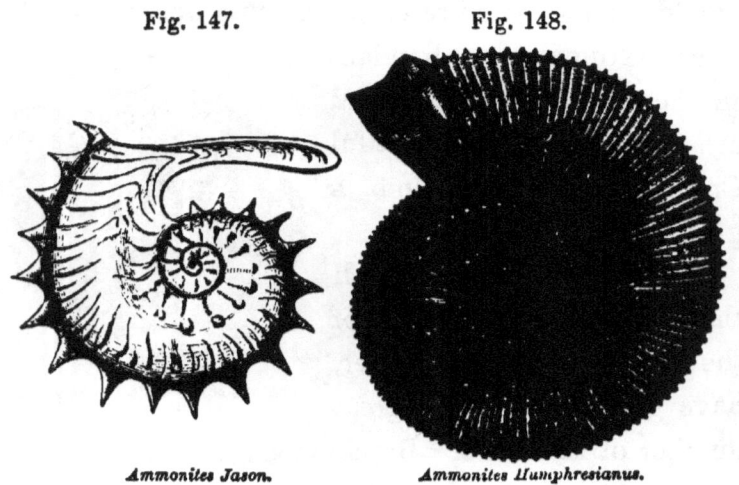

Fig. 147.   Fig. 148.
Ammonites Jason.   *Ammonites Humphresianus.*
Oolitic Cephalopods.

ARTICULATES are represented by several families of INSECTS.

VERTEBRATES appear in the form of Fishes, and Reptiles; and traces of the lowest order of Mammals occur in some of the upper stages of the system. But the true Mammals do not make their appearance till the Tertiary.

FISHES. Agassiz has described sixty species from the Lias of Lyme Regis, in England, not one species of which exists in our present waters.

REPTILES of this period, and the next, appear in

178   OOLITIC SYSTEM.

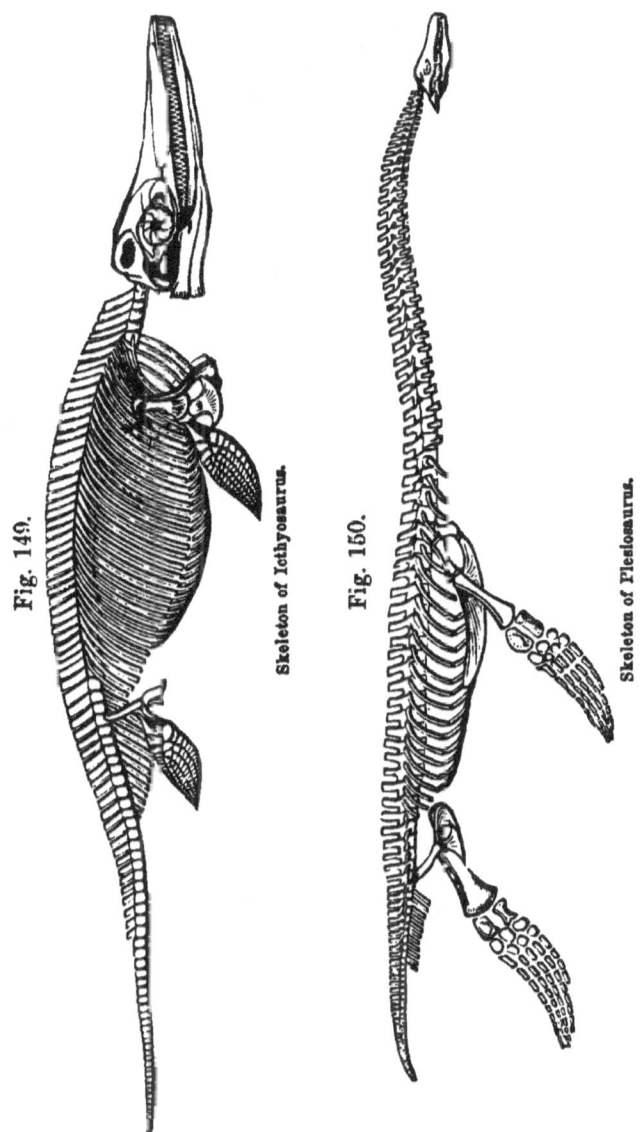

Fig. 149. Skeleton of Icthyosaurus.

Fig. 150. Skeleton of Plesiosaurus.

forms more strange and monstrous than those of any other age of the world. Some of the most extraordinary are known among naturalists, under the generic names of Icthyosaurus, Plesiosaurus, Megalosaurus, and Pterodactyl.

The Icthyosaurus was, in some instances, nearly 30 feet long. The skeleton of one in the British Museum, shows that the animal must have been 25 feet in length. In some specimens, the jaws are 6 feet long; the cavity for the eye 14 inches; the teeth 180 in number; and the vertebræ more than 100. The locomotive organs of this animal were paddles each composed of more than 100 bones. Its skin was naked, as shown by a portion found fossil. Its food consisted of fishes and reptiles, as shown by their half-digested remains found with the skeleton.

This animal is found both in England and on the Continent.

The Plesiosaurus was 11 feet long, as shown by a specimen in the British Museum. Its most remarkable peculiarity is its long neck, which contains 30 vertebræ. Twenty species have been found.

The Pterodactyl is the most wonderful reptile of this period, or any other. Its body was like that of a mammal, its wings like those of a bat, and its jaws and teeth like those of a crocodile. Probably this animal could walk, fly, or swim, as necessity re-

180 OOLITIC SYSTEM.

quired. Seven species have been found at Solenhofen, Bavaria.

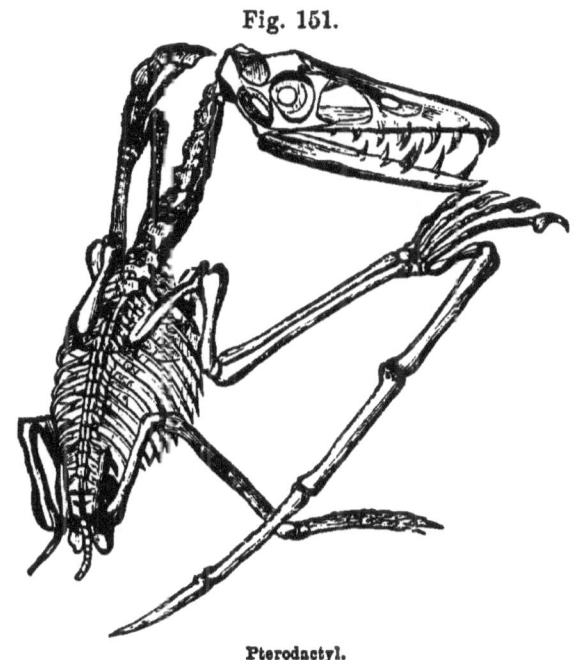

Fig. 151.

Pterodactyl.

It is in this period and the next that we find the first remains of genuine Chelonians.

SECTION VI.

CRETACEOUS, OR CHALK SYSTEM.

This system is extensively developed both in Europe and in North America, though in the latter the pure chalk beds are wanting.

It occupies a considerable portion of New Jersey, and thence appears at intervals as far south as Alabama; from the latter state it extends, in a widening belt, to the Rocky Mountains. In Europe, it extends from Great Britain, with some interruptions, across the continent to the southern part of Russia, and from Sweden to Bordeaux in France.

The rocks of this system are chalk and other calcareous deposits, sands, sandstones, and marls. The pure chalk strata are confined to the upper part of the system. Many of the beds of argillaceous and silicious limestones, also sands and sandstones, are of a green color; hence the term Greensand is often applied to a portion of the Cretaceous. Below the Greensand in the south-eastern part of England, there is a fresh-water deposit, called the Wealden, which we may class with the rocks of this period, though generally described as a distinct formation.

Fossil plants of the Cretaceous period are not very

abundant. Sponges and other low forms of vegetation are however often found in the flint nodules. But the rocks of this group abound with animal remains. In fact, the chalk-beds are almost wholly of animal origin.

RADIATES are numerous in the form of Corals and Echinoderms.

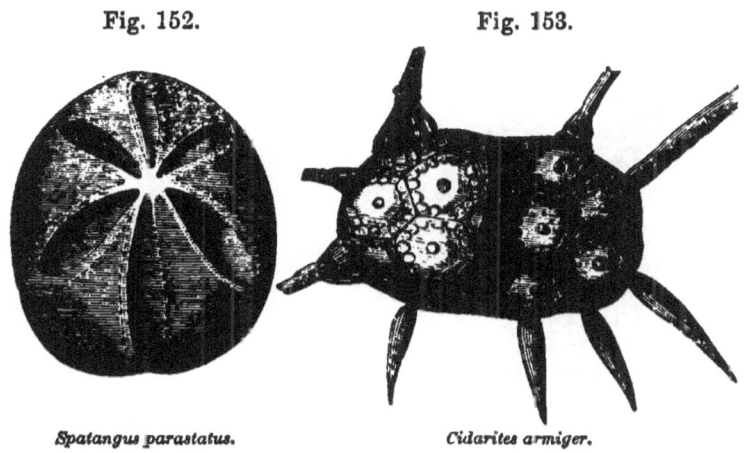

Fig. 152.   Fig. 153.

Spatangus parastatus.   *Cidarites armiger.*

Cretaceous Echinoderms.

MOLLUSCS are abundant, and appear in many new species.

The CEPHALOPODS are especially rich in variety of form. Fig. 154 represents a species of Belemnite common to the Cretaceous strata of the United States, Europe, and Asia. Fig. 156 reminds us of the Orthoceratites of the previous periods, but differs essentially from them in having dentated partitions, as

CRETACEOUS SYSTEM.

seen in the figure before us. This species is common in New Jersey, Alabama, and Texas.

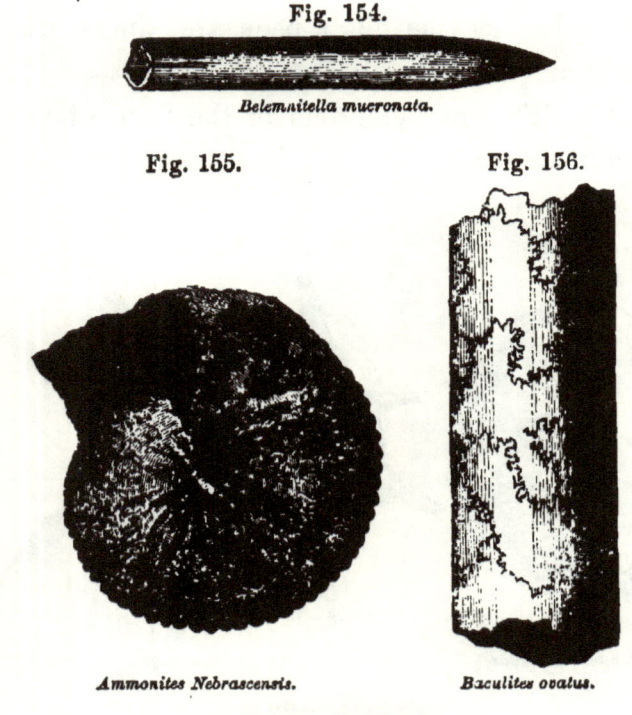

Fig. 154.
Belemnitella mucronata.

Fig. 155.
Ammonites Nebrascensis.

Fig. 156.
Baculites ovatus.

Cretaceous Cephalopods.

ARTICULATES are not numerous; but CRUSTACEANS are found in the chalk of Europe.

VERTEBRATES are represented by numerous Fishes and Reptiles. Birds have also been found in the Wealden.

FISHES are abundantly represented by teeth, jaws, and vertebræ; whole specimens being more rarely found than in rocks of a more compact nature. In this period Ctenoids and Cycloids first make their appearance upon the earth. Up to this era all the fishes, though exceedingly numerous, belong to the Placoids and Ganoids.

The REPTILES of this era, although of gigantic dimensions, seem more nearly related to those of the present day than any of those belonging to the preceding periods.

The *Mososaurus*, found at Mæstricht, and now in the Museum at Paris, was a reptile 25 feet long.

The *Iguanodon*, whose remains have been found in the Wealden, was a reptile 30 feet long, 14 feet in girth, and with a foot 6 feet in length. In the form of its teeth, this reptile resembled the modern Iguana of the tropical regions of America. Dr. Mantell, who has examined the bones of more than seventy individuals, estimates the length of a full-grown Iguanodon at 50 or 60 feet.

Such were the Reptiles that have left their skeletons to give us some idea of reptilian life during this and the preceding era.

The reign of gigantic reptiles terminated with the Cretaceous period.

SECTION VII.

TERTIARY SYSTEM.

This system consists of three distinct stages—the Lower, Middle, and Upper, or the Eocene, Miocene, and Pliocene.

The Tertiary system covers a large portion of the sea-board States, from New Jersey to Texas inclusive. It also occupies wide areas west of the Rocky Mountains. Tertiary deposits cover a large portion of Western, Central, and Southern Europe. London, Paris, and Vienna stand upon the Tertiary. It is also largely represented in South America and in Asia.

The rocks of this period are limestones, sandstones, gypsum, marls, and clays. Volcanic rocks also penetrate, and are interstratified with the rocks of this era. This is especially the case in Auvergne, France, where there are numerous craters of extinct volcanoes of Tertiary age. The interstratification of volcanic products with the sedimentary deposits, shows us that, in the districts above mentioned, volcanoes were very active during this period.

Beds of brown coal, of Tertiary age, are numerous both in this country and in Europe.

186   TERTIARY SYSTEM.

The Tertiary marks a new and important era in the earth's history. The existence of the present races of plants and animals is strikingly foreshadowed in its fossil organisms.

Plants are very abundant, especially in connection with the beds of brown coal and lignite.

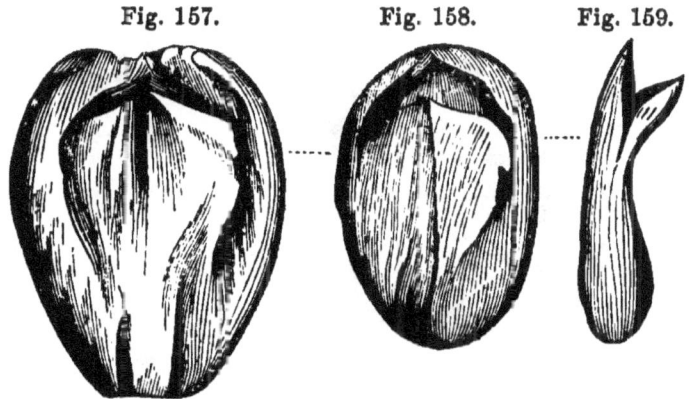

Fig. 157.   Fig. 158.   Fig. 159.

Fossil Fruit, Brandon, Vt.—Hitchcock.

Animals of this period are land, fresh-water, and marine, and are closely allied to those of the present day; belonging in most cases to the same families, and often to the same genera, and in some instances, in the case of shells, to the same species.

Species are also more limited in their geographical range than at any previous period, which further indicates an approach to the existing state of things on our globe.

RADIATES are abundant, but many of the forms common in the past ages are unknown in this. *Corals* are numerous, and are closely allied to those of the present day. Figs. 160 and 161 represent two species from the Tertiary of the Middle and Southern States.

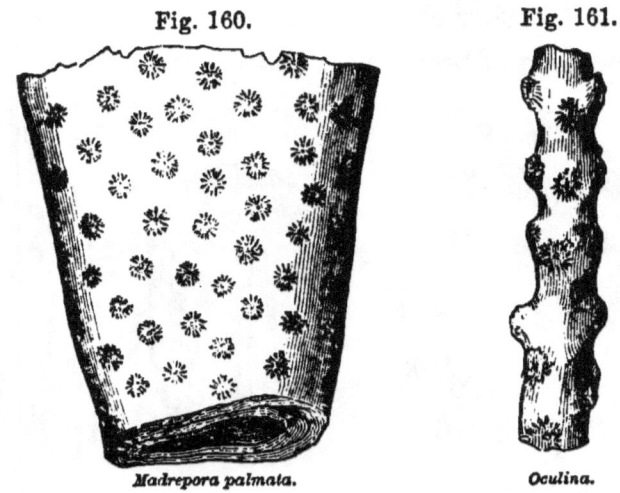

Fig. 160. *Madrepora palmata.* Fig. 161. *Oculina.*

Tertiary Corals.

Crinoids, which were exceedingly numerous in the paleozoic period, and even in the secondary, are here represented by only a few species. But *Asteroids* and *Echinoids* appear in great numbers, and in new forms.

MOLLUSCS. The absence of certain representatives of this branch is a striking evidence of a new geological era. The Ammonites and the Belemnites, which were so abundant in the last two systems, are not found in this. They ceased to exist at the close of the Cretaceous. Molluscs, however, are very abundantly represented, and by new species.

188 TERTIARY SYSTEM.

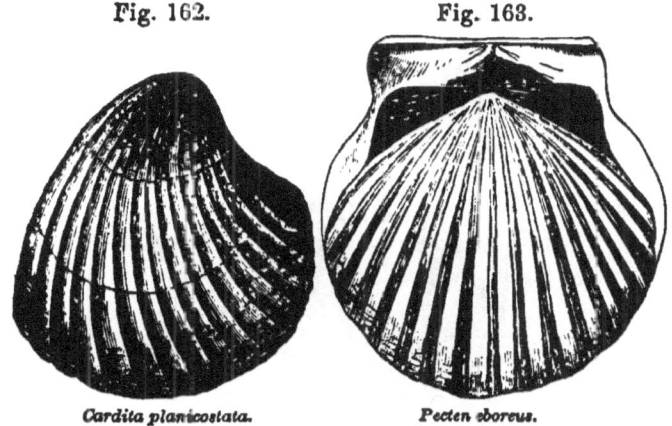

Fig. 162. Fig. 163.

Cardita planicostata.   *Pecten eboreus.*

Tertiary Acephals.

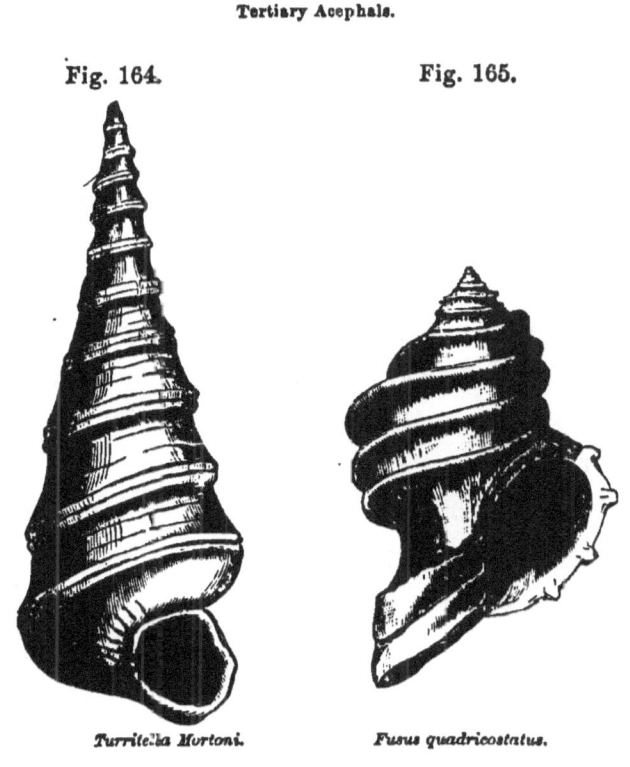

Fig. 164. Fig. 165.

Turritella Mortoni.   *Fusus quadricostatus.*

Tertiary Gasteropods.

Extensive beds of this era are mainly composed of small shells, many of which are microscopic. They are known under the name of Foraminifera. One

Fig. 166.

Nummulitic Limestone.

group of these animals has been called Nummulites, from their resemblance to a coin. Nummulitic rocks occur in Alabama, and they abound in the Alps and Pyrenees, also in Egypt, and in the Himalaya Mountains. Some of the pyramids of Egypt are built of Nummulitic limestone.

ARTICULATES are not abundantly preserved, but INSECTS and CRUSTACEANS are found to some extent.

VERTEBRATES. It is in this branch that the fauna of the Tertiary receives its most marked characteristics. In this system we find animals higher in rank than are found in any of the previous ones.

FISHES are numerous in the Tertiary. In the earlier periods, they were covered with a sort of enamel, but in this we find them covered with horny scales like the fishes of the present day. In the United States the teeth of sharks are abundant in the beds of this era.

Fig. 167.

Platax altissimus.

Tertiary Fish.

REPTILES, which especially characterize the Oölitic and Cretaceous systems, no longer stand in the first rank. Although numerous, they are not on that gigantic scale exhibited in the preceding periods.

BIRDS are found in the Tertiary beds around the city of Paris.

MAMMALS. In the Eocene of Clarke county, Ala., the remains of a gigantic Cetacean have been discovered. The bones occur in great numbers.

But the important fact to be here observed is this—the Tertiary marks the introduction of the true placental mammals. The sea and the estuaries, though rich in animal life, no longer furnish the highest representatives of the Animal Kingdom; but, in this period, land animals assume the first rank, which they ever afterwards maintain.

Fossil mammals are numerous in the Tertiary deposits of Nebraska. Twenty species of mammals have been found in the Eocene rocks of the Paris basin. These, as well as those of Nebraska, belong mostly to the order of Pachyderms. The careful researches of Cuvier enabled him to restore, to a great extent, the fauna of the earliest days of the Tertiary, as it existed in the vicinity of the present city of Paris; and he showed that it was composed of animals not only specifically but generically distinct from any of those now living.

One gigantic mammal of the Tertiary, Middle portion, has been named Dinotherium Its remains occur in Bavaria, Austria, and France. This animal was about 18 feet in length, and particularly remarkable on account of two tusks which turned downwards, and which were undoubtedly used in tearing up the roots of aquatic plants. It probably lived about the water, like the hippopotamus of the present time.

SECTION VIII.

DRIFT, OR BOULDER PERIOD.

The records of this era are spread out before us, over North America, north of parallel 40°, and over all the northern countries of Europe.

The Drift consists of materials derived from all the previous formations. These materials are in all stages, from the finest sand to boulders and fragments of rock of enormous dimensions, and, in many cases, all mixed together in confusion.

In some places the Drift material forms only a slight covering over the solid rocks; in some others it is piled up in ridges, and hills of great height, which may be found in all the northern parts of our country.

The boulders which fill and cover the soil, and those scattered upon the hills and mountains, belong to this formation, and, with the grooved and polished rocks, to be noticed hereafter, form its chief characteristics.

By examination, these boulders are found to be entirely different in structure and mineral composition from the rocks upon which they rest, and therefore are not in the places where they were formed, but have been transported to their present position by some powerful agency.

Everywhere in North America, within the limits above pointed out, the loose rocks have been moved from the north towards the south; and the ledge whence a boulder was derived, can, in most cases, be found to the north of the present position of the boulder, but never south of it. In some cases they have been carried only a few rods from the parent ledge, in others, many miles.

Long Island, Martha's Vineyard, Nantucket, and other islands on our coast, are covered with boulders derived from the ledges of the continent. Many parts of Cape Cod, where there are no ledges, are covered with boulders of all sizes, and those widely different from one another in mineral composition.

Boulders of porphyritic iron ore, from Iron Hill, Cumberland, R. I., are scattered over all the regions of the state south of that locality. This is a very marked case; as the iron ore is so well characterized, that there is no difficulty in identifying the boulders with the ore in place at Iron Hill. Boulders derived from the country north of the Great Lakes are scattered over the Western States.

In numerous instances the rocks from one range of hills or mountains, have been carried across deep valleys, and scattered upon opposite mountains, and the country beyond. On Hoosac Mountain, Mass., there is a boulder one thousand feet above the valley across which it has been transported.

It is common to find boulders having the outside covered with grooves, parallel to one another. The writer has noticed many perfect examples of this kind in Lancaster, Mass. These grooves have undoubtedly resulted from the grating of the boulders over ledges, while the former were fixed in the ice, which, as will be shown hereafter, was concerned in their transport. Boulders are often found so nicely poised on other

Fig. 168.

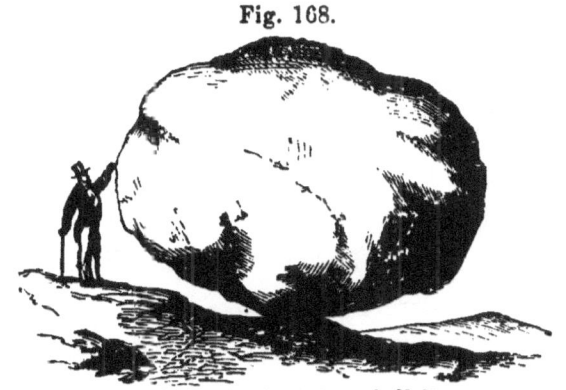

Rocking Stone, South Acworth, N. H.

rocks, and on the mountain sides, or summits, that they may be moved with the hand, though it would require immense power to dislodge them. These are called *Rocking Stones.*

Throughout the regions already pointed out as occupied by the Drift materials, the rocks in place are more or less polished, striated, or grooved. These marks are found on all the consolidated formations that appear at the surface, and they constitute an essential part of the records of the Drift period.

Wherever the rocks have not suffered decomposition or disintegration, these markings are conspicuous in the valleys, on the hills, and on the mountains. The grooves are generally a fraction of an inch in depth, and from a quarter of an inch to three or four inches in width. Rarely, the grooves or furrows occur a foot or more wide, and several inches deep. Again, the markings are so finely cut as to be scarcely visible; and sometimes, as on quartz and other hard rocks, a lens reveals the most definitely-cut striæ where none were to be seen with the naked eye. Sometimes the rocks are completely polished. There is every grade of these phenomena to be seen in the Drift regions, and not unfrequently all the varieties mentioned above are found at one locality.

The general direction of the striæ, or grooves, is the same as that in which the boulders have been transported; that is, from north to south, varying from a few degrees west of north, and east of south, to a few degrees east of north, and west of south.

The markings are more numerous on the northern flanks of the mountains than on the southern; and the northern portions of mountains everywhere in the Drift regions of North America, are rounded, while the southern portions are more angular.

These facts, as well as those before noticed, show that the agency which transported the boulders and other drift material, and which furrowed, rounded, and polished the rocks, operated from the north towards the south.

As these phenomena of the grooved and polished rocks may be seen everywhere within the regions named above, we need point out only a few cases.

There are good examples near the city of Portland, Me., especially on the rocks near the sea-side, where the striæ are distinctly cut, running parallel and for great distances, and disappearing beneath the sea. The writer has also noticed fine examples throughout almost the entire width of the same state.

The mica slate, gneissoid, and other hard rocks, everywhere furnish good examples. Fig. 169 represents a fragment of striated and polished quartz from Williams Hill, the celebrated beryl locality in Acworth, N. H. The whole top of the hill is rounded and smoothed—a result of the Drift action. The White Mountain range, in all favorable places, exhibits scratched and furrowed rocks. Mt. Chocorua is furrowed from the base to the summit, and in many instances the grooves are very deep.

Fig. 169.

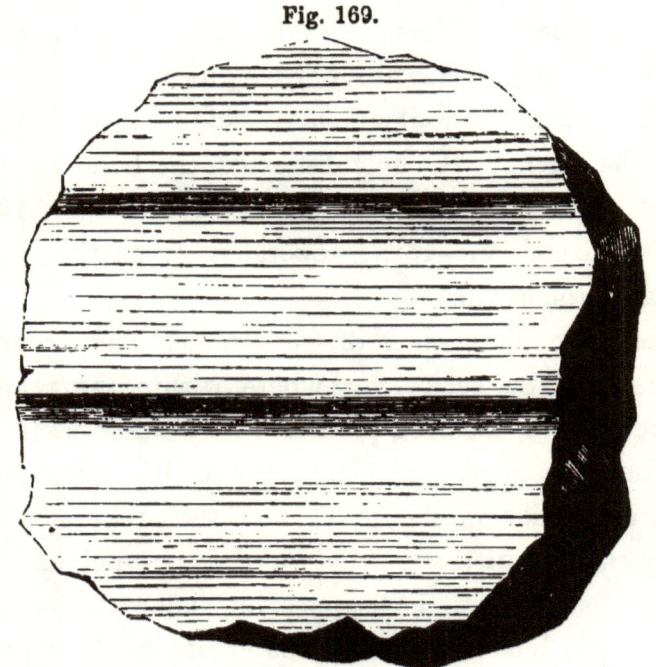

Fragment of Striated Quartz, Williams Hill, Acworth, N. H.

Fig. 170 represents a polished and finely striated fragment of quartz, broken from a vein of that material on the summit of Mt. Clinton, one of the White Mountains. The portion left white in the figure, is higher than the other parts of the specimen, and is polished as smooth as a pane of window-glass.

This smooth quartz shows that the ice which passed over this point, and thus polished it, rested immediately upon it, with only sand, ground as fine as dust, between.

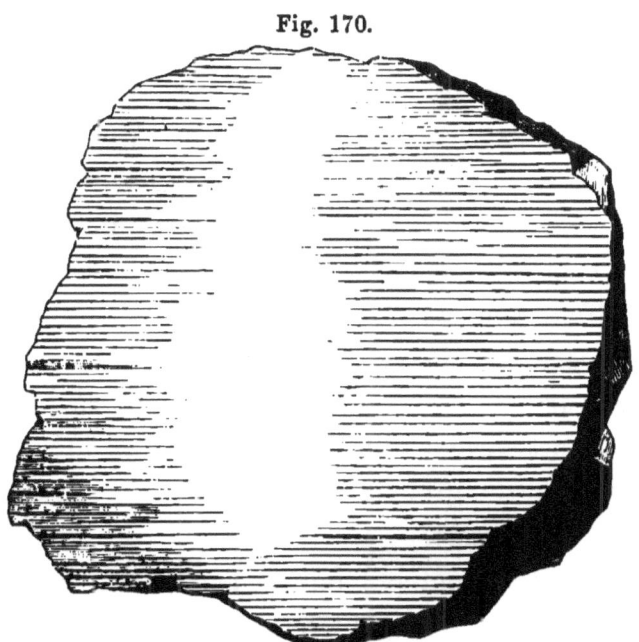

Fig. 170.

Polished Quartz, Mt. Clinton, N. H.

The writer thinks he has detected distinct grooves on Mt. Washington, some distance above the Lake of the Clouds. These grooves are near the bridle path which is followed in ascending this mountain from the "Notch," and probably are at the highest point east of the Rocky Mountains, where evidences of the Drift period have been observed. It is probable that the Drift agency did not extend to the summit of Mt. Washington, as that is at present covered with angular fragments and blocks of the peculiar mica slate of which that mountain is mainly composed. Had this agency swept over the summit, it is believed we should find evidences of the fact in grooves, and in the absence of angular blocks, which are now so abundant. But we must not forget that time enough has elapsed, since the period of the Drift, to allow the frost to break up the summit into fragments, so as to give it the present appearance, though it may have been swept over in the same manner as other parts of the Drift region.

The rocks of Massachusetts are covered with well-defined striæ. Professor Hitchcock, who has studied this subject very extensively, has pointed out, and mapped out, the grooves and their directions throughout the entire state.

A perfect example of polished and grooved rocks, may be seen a few rods from the Railroad Station in Greenfield, Mass. The rocks on the coast of this state are covered with striæ. Visitors to Nahant can see beautiful examples near the summer residence of Agassiz.

The lamented and learned naturalist, Professor C. B. Adams, observed more than three hundred distinct cases in Vermont. He mentions the Valley of Winooski, as especially interesting on account of its numerous polished and grooved rocks; some of them exhibiting furrows three or four inches deep, and twelve to thirty inches wide.

At Rochester, New York, the limestone, over a wide area, is polished and striated in the most perfect manner.

I have specified only a few localities. Want of room forbids me to mention more. The student must not regard these as the most interesting, or important. Undoubtedly there are hundreds, and probably thousands, of cases in every state occupied by the Drift, as interesting, and as prominently marked as those pointed out above.

The remark made in the outset is perfectly true, that these markings, and other Drift phenomena, are found in every part of the northern portion of our country, at all levels, to the height of 5000 or 6000 feet above the sea.

An explanation of the Drift phenomena is not an easy task, and probably will not be fully accomplished, until there shall be a more complete survey of all the facts which are within our reach. Some points, however, are well settled. The straight and parallel striæ, or furrows on the rocks, could not have been produced by mere currents of water, which are deflected from their course by every obstacle; while these markings show that the agency moved equally well over all kinds of surface, ploughing through, and grinding down whatever impeded its progress, and turned aside only by the most formidable obstruction, such as a lofty mountain.

Mere currents of water never could have lifted boulders from lower to higher levels, or transported them from one range of mountains, across deep valleys, to another range, and to the country beyond.

Ice must have been concerned in producing these phenomena. It has moved over the whole Drift region, either in the form of Glaciers, or Icebergs, and perhaps both and with rocks and pebbles frozen into the lower surface, has acted like a huge rasp to wear away, furrow, and polish the hardest rocks, and, at the same time, to bear forward boulders, fragments of rock, and other material, and leave them far from their original places.

Further evidence that the Drift phenomena were produced by Icebergs, Glaciers, or both, will be introduced in a subsequent chapter, where geological changes will be noticed at some length.

SECTION IX.

ALLUVIUM.

Under this head we may include all the accumulations newer than the Drift, described in the last section. Alluvium and Drift so blend that it is almost impossible to draw a line of distinction between them; and Professor Hitchcock, and some other eminent geologists, comprise under Alluvium all accumulations newer than the Tertiary, considering Drift only as the first part of one great geological system; and it will be recollected that the two are grouped together in our table showing the classification of rocks.

There is, however, one general distinction which may be made between Drift proper and Alluvium, according to our limitation;—Drift is not stratified; the sand, gravel, and boulders, are all piled together in confusion, while accumulations newer than the Drift are stratified, or exhibit more or less of a sorting of the materials.

Under Alluvium we include, then, ancient sea beaches, rounded hills of pebbles, lake and river terraces, deltas, deposits of sands and clays; also, accumulations of marls, peat, calcareous tufa, bog iron ore, &c.,—embracing all the progressive accumulations.

At the close of the Drift Period proper, or the period described in the last section, the Drift regions were covered with a mass of sand, gravel, rounded and angular boulders confusedly mixed together—the result of the Drift operations. These materials have been more or less worked over and modified by the ocean, lakes, and rivers, giving rise to the varied forms of Alluvium enumerated above,—excepting only the peat, marls, tufa, &c., which are formed by processes to be explained hereafter.

Professor Hitchcock, whose researches in the Surface Geology of this country have probably been more extensive than those of any other man, and whose interesting published results may be found in the 9th volume of the "Smithsonian Contributions to Knowledge," finds ancient sea beaches in many parts of New England, and at all levels to the height of 2000 or 3000 feet.

The rounded hills of pebbles that border many of our mountains are but drift material worked over by the waves of an ancient sea. Such modified materials, and in fact all the altered Drift, are called *Modified Drift*.

All are familiar with the fact that along our rivers there are generally several terraces, one above the other, each higher one being farther back from the

ALLUVIUM.

Fig. 171.

Hills of Modified Drift.

river than the one below it. Sometimes these terraces appear on both sides of the stream, and sometimes only on one side. The number varies from two or three to ten. The highest or first-formed, is often at a great height above the present channel; but at whatever height we find the terrace, we may be sure that it was once the bank of the river.

These terraces are the result of the natural drainage of the continents. At the last emergence of the lands from the ocean, the rivers all ran over the surface of the loose material through which they have ever since been cutting down their channels, and leaving terrace after terrace, as evidences of their operations.

Fine examples of river terraces may be seen in nearly all parts of the United States and especially along the rivers of New England. In many cases they may readily be traced as one rides in the rail-car. This is especially true of those along the Merrimac

ALLUVIUM. 205

river, which are very conspicuous between Lowell and Concord.

But the localities where good terraces may be seen are so numerous that it is not necessary to point out particular cases. Fig. 172, reduced from Professor Hitchcock's paper in the Smithsonian Contributions, gives a good idea of the form and position of river terraces.

Fig. 172.

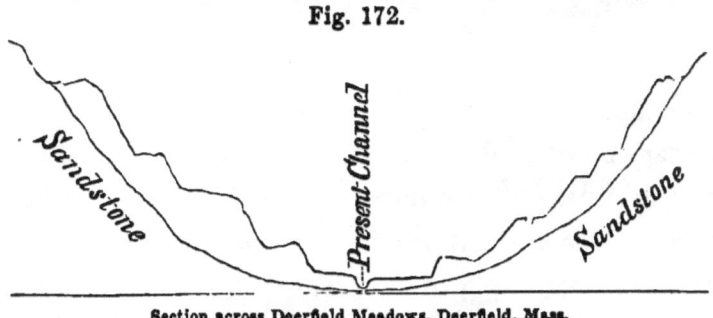

Section across Deerfield Meadows, Deerfield, Mass.

The deltas, marls, peat, &c., of this period will be sufficiently noticed in a subsequent chapter, and therefore need not be spoken of here.

In the clay deposits of this era it is common to find nodules, or concretions of argillaceous matter, called Clay-stones. They are probably formed by the affinity that particles have for one another. Not unfrequently the nucleus is a scale of a fish, or a shell, or some other organic substance.

Clay-stones occur of almost every variety of form, sometimes taking shapes which, with the aid of a little imagination, appear to be those of familiar animals,

206 ALLUVIUM.

as a cat, dog, hare, butterfly, &c. Professor Hitchcock has figured one in his Elementary Geology, which bears a striking resemblance to the human head. The accompanying figures are exact representations of two specimens from East Windsor, Conn., presented to the writer by Rev. E. H. Pratt.

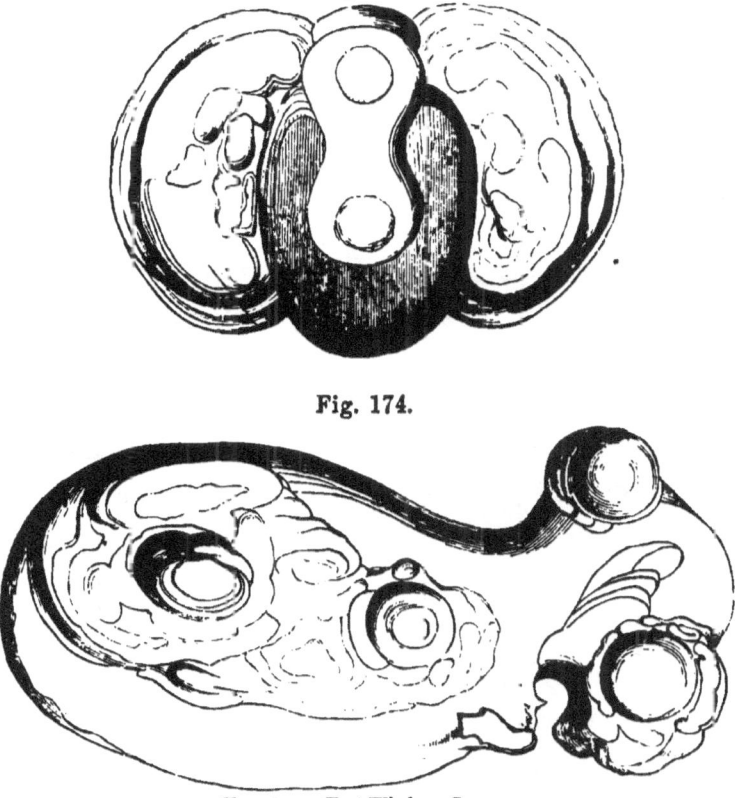

Fig. 173.

Fig. 174.

Clay-stones, East Windsor, Conn.

In the Drift proper there are few or no fossils, but in the deposits newer than the Drift they are very abundant and full of interest. The BIRDS and MAMMALS of this era especially claim our attention.

In New Zealand, bones of a bird, called Dinornis, have been found which exceed in bulk those of the largest horse. The thigh bone of this bird is sixteen inches in length, and nine inches in circumference in the middle part; and the upper part of the tibia measures twenty-one inches in circumference. This bird, when alive, was eleven or twelve feet high; as tall as the largest elephant. These bones are now in the Museum of the College of Surgeons, London.

In Siberia the unconsolidated deposits so abound with elephant's tusks, that for years the ivory secured there, has constituted a regular article of commerce. This fact shows us that Siberia was once a land of elephants.

Near the close of the last century, an elephant was found on the northern coast of Siberia, imbedded in the ice. It was in such a perfect state of preservation, that the people of the vicinity fed their dogs upon its flesh. This animal was covered with hair between one and two inches long, and was evidently adapted to live in a cool climate. How this elephant became imbedded there, in so perfect a state of preservation, is, perhaps, a question yet to be solved.

In Great Britain, and other parts of Europe, the caves and the superficial deposits contain remains of forms that are no longer represented in that quarter of the globe. These remains show us that Great Britain was once inhabited by at least two species of elephant, two species of rhinoceros, the hippopotamus, tiger, three species of bears, hyenas, and a gigantic elk, all of which have passed away.

The great Irish Elk was found in the shell marl below the peat, and is ten feet high to the top of its horns, whose tips are ten feet apart. Our American Elk appears small beside this gigantic Ruminant of other days. The skeleton of this animal is in the Museum of the College of Surgeons, mentioned above.

But in no country are the remains of this era more interesting than in the United States; where, besides the remains of a great number of smaller Mammals, we find those of the Mylodon, Megatherium, and Mastodon, huge animals that no longer live on the globe. And the researches of Professor Holmes, in the superficial deposits of South Carolina, whence he is revealing wonders, have brought to light the remains of an extinct horse. It is highly probable that this noble animal was indigenous to this country, and became extinct before the arrival of the white man. Our present horses are descendants of those introduced from abroad.

The Megatherium was at least twelve feet long, and eight feet high. Its thigh bone is three times as thick as that of the elephant, and the spinal cavity indicates a spinal cord a foot in diameter! The remains of

ALLUVIUM. 209

this animal have been found in South Carolina and Georgia; also abundantly in South America.

Fig. 175.

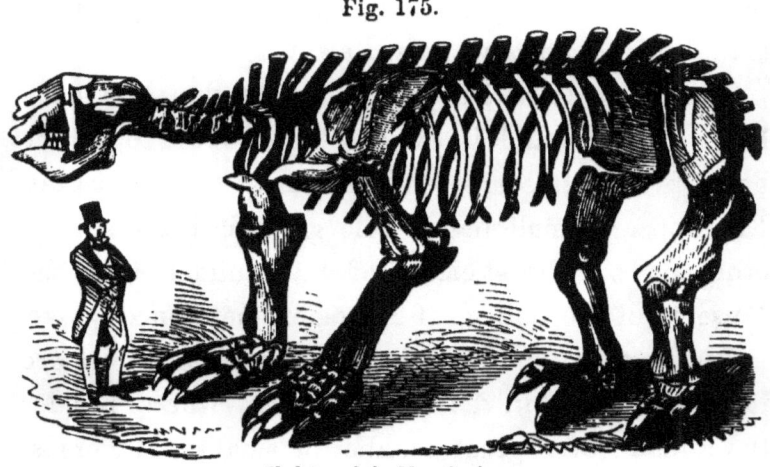

Skeleton of the Megatherium.

The remains of the Mastodon have been found in great abundance in the United States, occurring almost everywhere west of the Connecticut river. This animal resembles the elephant, but is entirely distinct from the latter. Merely a tooth is enough to distinguish these animals from one another, as the grinding surface of the tooth of the Mastodon is composed of conical projections, and that of the elephant is flat, as may be seen by the figures on the next page.

The most perfect skeleton of the Mastodon in the United States, is now in the private Museum of the late Dr. John C. Warren of Boston, who published a

210 ALLUVIUM.

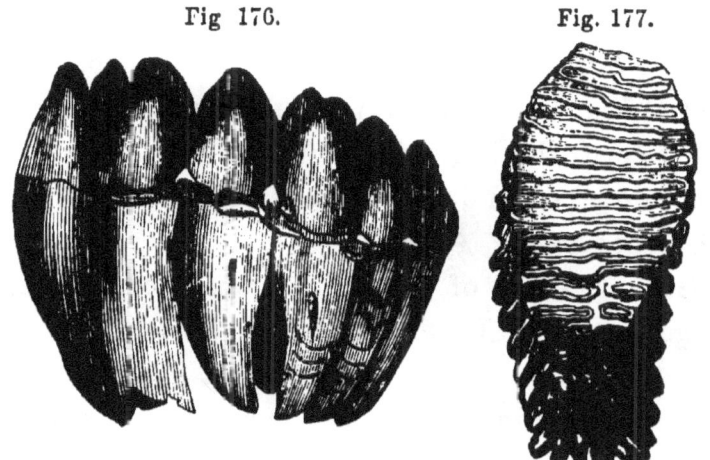

Fig 176.

Fig. 177.

Molar of the Mastodon—side view.

Molar of the Elephant—grinding surface inclined towards the observer.

large volume wholly devoted to a description of this ancient Pachyderm of our country.

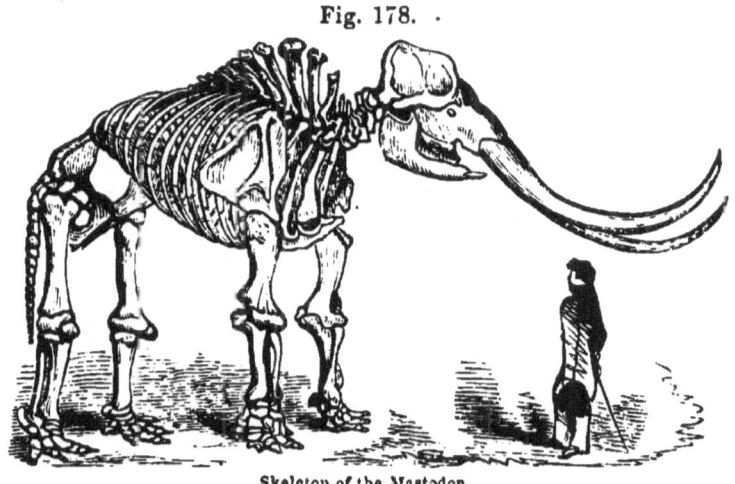

Fig. 178.

Skeleton of the Mastodon.

This skeleton was discovered in 1845, in marl

below a peat bog, in Newburgh, N. Y., 70 miles from New York city. The height of the skeleton is 11 feet. The length from the anterior extremity of the face to the beginning of the tail 17 feet, and the length of the tail 6 feet. The tusks are each 10 feet 11 inches long.

Such was one of the animals once common in a large part of the United States.

Neither the remains of man, nor any of his works, have been found in any formation below Alluvium, nor are they found in the lowest portions of this. Geology, as well as Revelation, thus shows that of all the Animal Kingdom, Man was the last to make his appearance upon the earth.

I have thus endeavored to present the leading facts revealed by the study of each geological system. It is now important that the student be able to survey the ground over which he has passed, and to retain in memory the results which he has reached; therefore the following Tables, showing the Geological Periods when the different Branches and Classes of the Vegetable and of the Animal Kingdom appeared upon the earth, will be of service to the beginner in geological science.

These tables represent the several periods in their chronological order, the lowest being oldest. The first Table relates to the introduction of Plants, and the second to that of Animals.

The dotted lines show when each Branch or Class was introduced, and the periods through which the same lived.

VEGETABLE KINGDOM.

	Thallophytes.	Acrogens.	Gymnosperms.	Monocotyledons.	Dicotyledons.
Alluvium and Drift . . .	:	:	:	:	:
Tertiary	:	:	:	:	:
Cretaceous	:	:	:	:	:
Oölitic	:	:	:	:	
New Red Sandstone . .	:	:	:	:	
Carboniferous	:	:	:	:	
Old Red Sandstone . . .	:	:	:		
Silurian	:	·			

No separate column is made for Anophytes, which are here included with the two lowest groups.

Thallophytes were introduced at the beginning of the Silurian; Acrogens later in the same period; Gymnosperms during the Old Red Sandstone; Monocotyledons in the Carboniferous; and true Dicotyledons at the beginning of the Cretaceous.

ANIMAL KINGDOM.

	Radiates.			Molluscs.			Articulates.			Vertebrates.					
	Polyps.	Acalephs.	Echinoderms.	Acephals.	Gasteropods.	Cephalopods.	Worms.	Crustaceans.	Insects.	Fishes.	Batrachians.	Reptiles.	Birds.	Mammals.	Man.
Alluvium and Drift															
Tertiary . . .															
Cretaceous . .															
Oölitic . . .															
New Red Sand.															
Carboniferous .															
Old Red Sand. .															
Silurian . . .															

It may be doubtful whether Reptiles and Birds appear quite as early as here indicated.

The student will please bear in mind that although each Class lived through all the periods after its introduction, it is represented in the successive periods by new *Species*, often by new *Genera*, and in many cases by new *Orders*.

By way of recapitulation, we may add, that the Paleozoic Period is especially characterized by its beautiful Crinoids, numerous and unique Brachiopods, its Orthoceratites, Trilobites, and Fishes; the Secondary, by its Ammonites, Belemnites, Nautili, and enormous and strange-formed Reptiles; the Tertiary and Modern, by numerous Bivalves, and Gasteropods, and gigantic Mammals.

In concluding this chapter, we may add, that the geological systems afford data by which the relative ages of mountain ranges and continents may be determined.

If Silurian strata form the flanks of a mountain chain, we know that these mountains were uplifted after the formation of the Silurian rocks, else these rocks would not have been upheaved.

If Tertiary rocks constitute the flanks, then the mountains were uplifted after the Tertiary period.

If Silurian beds flank a mountain range, and undisturbed Tertiary strata occur in connection with these beds, we know that the range of mountains was uplifted after the Silurian, but before the Tertiary; for had the Tertiary beds been formed before the upheaval, they too would have been uplifted.

By the application of such principles, geologists have found out, that not only the continents, but different parts of the same continent, differ greatly in age.

Geologically considered, America is the Old World. A large part of North America was dry land, while yet a large portion of Europe formed the bottom of the ocean.

Geologists have also found out that the highest mountains of every country are the newest or last elevated. The Alps are younger than the mountains of Scandinavia; the Rocky Mountains younger than the Alleghanies, and the latter younger than the White Hills of New Hampshire.

CHAPTER IX.

THE GEOLOGICAL CHANGES NOW GOING ON, AND THE AGENCIES BY WHICH THEY ARE PRODUCED.

. IN our previous investigations, we have learned that the earth has undergone great changes; and if we look at the earth now, we find nothing stationary. Probably our planet does not present a single feature that it possessed when spoken into existence by its Great Author.

We are accustomed to look upon the continents, the mountains, and the hills as fixed; the courses of rivers, the barriers of lakes, seas, and oceans, as permanent; but all these are modified every century, every year, and every hour.

The agencies that are producing geological changes, are mainly Chemical, Aqueous, Igneous, and Organic.

Besides these agencies, the winds also produce great changes in many parts of the world, by drifting the sands.

On account of the affinity one element has for another, bodies are constantly decomposing, and their elements enter into new relations, and appear in new forms. Chemical agencies are slow in their operations; but their results are great. The operations of the other three classes of agencies are more open to view, and will be spoken of somewhat at length, in the three following sections.

SECTION I.

AQUEOUS AGENCIES.

Aqueous agencies operate in the form of Rains, Frost, Springs, Rivers, Waves, Currents, Glaciers, and Icebergs.

RAINS are both chemical and mechanical in their operations. In falling through the atmosphere, water absorbs more or less of carbonic acid, by the aid of which it decomposes all rocks containing lime, and causes many other chemical changes, by bringing in contact substances of different kinds. There is no shower that does not wash some of the loose material from the hills and mountains, into the valleys.

FROST is a powerful modifying agent. Water finds its way into the fissures of the rocks, where it freezes, and rends the rocks asunder, and the sides of the mountains are covered with fragments of all sizes thus produced. Immense boulders are often rent by frost, and cause much curious speculation by those ignorant of this fact.

SPRINGS produce changes by depositing substances which they hold in solution, such as lime, and iron; thus forming beds of calcareous tufa, and bog iron ore.

RIVERS cut channels in the superficial accumulations, and through the solid rocks, and transport loose material into the water basins.

Rivers have excavated deep channels in the rocks in every part of our country, giving rise to varied and beautiful scenery. These excavations afford the geologist an opportunity of studying the strata, and their contents, to the best advantage.

At Trenton Falls, N. Y., Canada Creek has cut its way in the Trenton limestone to a depth of 100 to 200 feet, revealing the strata that accumulated on the bottom of an ancient ocean, and which are now filled with the remains of life.

This is one of the most interesting places which the tourist and geologist can visit in our country. The deep ravines, the numerous waterfalls and cascades, the perfectly defined strata filled with trilobites and other fossils, richly reward the visitor.

The Genesee river, N. Y., has cut a channel in some parts of its course, 300 feet deep, the depth varying with the nature of the rocks. This river shows well how the different kinds of strata cause a series of falls.

The hardest rock is at the fall marked 1, the next in hardness at the one marked 2, and the softest at 3.

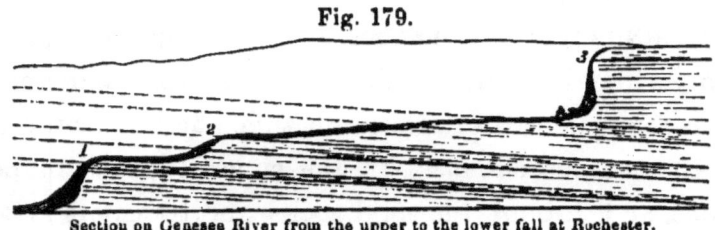

Fig. 179.

Section on Genesee River from the upper to the lower fall at Rochester.

Niagara river has cut a channel through the solid rocks, 200 feet deep, 1200 to 2000 feet wide, and 7 miles long. The evidence is conclusive that the Falls were formerly at Queenstown, seven miles below their present situation.

The high bluff at Queenstown, and the sides of the deep chasm, afford a fine exhibition of the different kinds of rocks in this region, and enable us to understand the present operations at the Falls, and the prospective results of these operations.

The lower rock is a soft shale, which rapidly wastes by the action of the water, while the overlying rock is a compact limestone. The result of this arrangement is, that the lower rock wears away, leaving the vast thickness of limestone overhanging, which is soon broken down by the 600,000 tons of water which pass over it each minute.

The "Table Rock," parts of which have fallen within a few years, is the upper limestone, and the cause of its falling is the destruction of the shales beneath, as before remarked

The opinion was formerly popular that the Falls would ultimately recede to Lake Erie, when that lake would be suddenly drained, and inundate the whole country below. But the dip of the rocks soon carries the shales so far below the bed of the river, that they will cease to be acted upon, while the hard limestone will occupy the whole height of the Falls. Then the rock being throughout of the same hardness, the abrupt fall will disappear, and a gradual descent take its place. So that, in process of time, there will be a gentle slope from Lake Erie to Ontario; and though the former be drained, it will be done gradually.

For the last fifty years, the Falls of Niagara have attracted the attention of the lover of sublime scenery, and the man of science. While both have feasted their souls on the grandeur and sublimity there exhibited, and which no pen or pencil can shadow forth, the latter especially has looked upon Niagara as one of nature's grand chronometers, registering the ages as they pass.

The rate at which the Falls recede cannot be definitely stated, but from the most reliable data that can be obtained, it has been shown that they have not receded more than one foot a year for the last half century. If this has been the rate of retrocession for the whole distance,—and on account of the nature of the rocks there is no reason for supposing it greater,—it has required 36,000 years for that great excavation. And that this work did not begin till after the Drift period, is proved by the remains of the Drift material which constituted the banks of the river before it began to operate on the solid rocks beneath.

Examples of river action are common where no river now exists. There is one case which the writer has recently examined with some care, and may be mentioned in this connection. In Bedford, Hillsboro' county, N. H., there are extensive excavations in the mica slate, where now, except in freshets, there is only a rill. At one place the channel, which has been cut in the solid rock, is 70 or 80 feet deep, and from 20 to 30 wide. The sides are waterworn, and one side, by its numerous broad grooves, shows very plainly the varying direction of the main force of the current in the successive steps of the process of excavation. At the head of the channel there is a large pool of water. In other parts the bottom is filled with rubbish through which, in one place, a pole may be passed down 20 or 30 feet. The cut on the next page, drawn from a sketch made on the spot by Professor Krüsi, gives a good idea of this place, which is one peculiarly interesting, and is much visited. It is known in the neighborhood as the "Devil's Pulpit." And here I wish to protest against the habit of dedicating to His Satanic Majesty some of the most interesting places in every region; and would suggest that this locality be hereafter called the *Bedford Ravine*.

Fig. 180.

Upper extremity of Bedford Ravine, Bedford, N. H.

A place called "Purgatory," in or near Mt. Vernon, N. H., presents us with results scarcely less interesting than those of the "Bedford Ravine;" and they show that ages must have elapsed since the process of excavation began, and that a much larger stream than the present, once flowed through that channel.

Old river channels of great depth are found in the western and south-western parts of North America.

Deep wells or "Pot-holes" are everywhere common along rapid brooks and rivers. Fine examples may be seen at Bellows Falls on the Connecticut, and at Amoskeag Falls on the Merrimac. Undoubtedly there are hundreds of places in our country that exhibit examples just as interesting.

Fig. 181.

The "Basin," Franconia Notch.

The celebrated "Basin" at Franconia Notch is one

of these wells, 40 feet in diameter, and 28 feet deep. It is filled to the depth of 8 or 10 feet with pure water, which revolves with such a force that it is considered a dangerous place for even an expert swimmer. No one can doubt for a moment that the entire basin has been made by the same process which is now going on there. How long the waters of the Pemigewasset have been revolving in this basin we may never know, but undoubtedly for ages.

But these pot-holes are abundant far above, and far distant from the present river channels, indicating, as well as the terraces before noticed, that the rivers have run at much higher levels. If the student examine carefully the ledges which form the sides of the valleys, he will find many of these wells at heights and distances from the present river bed that will astonish him.

On Union river, in Maine, the writer has noticed pot-holes 3 or 10 feet deep, and 50 to 100 feet above the present bed of the river; and what is seen on this stream may be observed on many others in the rocky portions of our country. Deep pot-holes occur both at Bedford Ravine and the place called Purgatory, noticed above.

But the most interesting case of this kind is found in Orange, N. H. Here, at an elevation of 1000 feet above the waters of the Connecticut and Merrimac, and 1200 feet above the ocean, there are several of these pot-holes in hard granite. One is 11 feet deep, and 4 feet in diameter. They were unquestionably produced by a waterfall—by the whirling of pebbles by the water; but when, it is not easy to say.

The rivers are carrying the lands into the sea; and the amount of material that has been transported by them, since the present order of things commenced, is truly wonderful.

A large part of Louisiana has been formed by material brought down by the Mississippi, and the land is still advancing into the Gulf of Mexico from the same cause. It is estimated that 28,000,000,000 cubic feet of sediment are annually carried down by this river, and deposited at its mouth, or swept away by the waves and currents.

The Amazon is so charged with sediment that its waters can be detected, by their discoloration, three hundred miles from its mouth. A part of the sediment brought down by this river is carried by the current of that region, and deposited on the coast of Guiana, which is thereby advancing upon the sea.

If we pass to the Eastern Continent, we find remarkable changes which have been produced by rivers. The Nile has formed a vast delta at its mouth. There is good evidence that nearly all Egypt is the gift of this river. The sediment brought down by the Nile, and deposited on the adjacent country, when the river overflows its banks, renders the soil productive.

P

Many of our valleys owe their fertility, as well as their gradual elevation, to their being annually inundated.

The amount of sediment carried down by the Ganges and Brahmapootra shows us how fast the rivers, in some parts of the world, are transporting the lands into the ocean. Sir Charles Lyell says that "if a fleet of eighty vessels, each freighted with 1400 tons weight of mud, were to sail down that river every hour of every day and night, for four months continuously, it would only transport from the higher regions to the sea, a quantity of matter equal to that carried by the Ganges in the four months of the flood season."

The Brahmapootra transports about the same quantity as the Ganges; and both these rivers, according to the distinguished author just mentioned, annually carry down to their delta forty thousand millions of cubic feet of sediment.

WAVES are producing geological changes. In one place they wear away coasts; in another, they pile hills of sand and pebbles as monuments of their power. They round the pebbles upon the shore, and their continued action reduces them to sand.

Waves produce rapid changes when they have access to a coast of loose materials, such as sand and gravel. Boston harbor undoubtedly was once filled with loose material, which has been removed by the waves; and the islands there and in the vicinity, are only remnants of a mass of land which has been separated from the continent. Nantucket and Martha's Vineyard have also been separated from the main land by the same agency.

The results of wave action, upon a given coast vary according to the nature of the rocks. Here the rocks yield quickly; there they are firm and compact, and stand out as bold headlands. Trap dikes are often removed by the waves; and thus long narrow channels are formed, extending far inland.

OCEAN CURRENTS are continually producing geological changes. Sweeping past every coast, they take a portion of the loose material brought down by the rivers, and bear it away, some of it hundreds of miles, and spread it out upon the floor of the ocean.

TIDES conspire with the waves and currents to modify the coasts, and transport loose material. Lakes and inland seas have no tides, or those only a few inches in height. The tides of the ocean vary according to the nature of the coast, from a few feet to 30 or 40 feet. In the Bay of Fundy the tide rises 70 feet.

In some situations a layer of sand or mud is deposited at each flow of the tide, and thus the successive layers retain, more or less perfectly, the impression made upon each, while it lay drying, during the time the tide was out.

We thus see processes going on in our bays and estuaries, which explain how the foot-prints and rain-drops of Connecticut valley were preserved. The unconsolidated deposits in the Bay of Fundy retain tracks and other impressions on the successive layers.

GLACIERS are vast masses of ice, of a peculiar structure, encased in high mountain valleys. They are, in fact, the transformed snow which falls upon the mountains above the snow line. This snow is gradually transformed into the glacier just as the snow upon the roof is transformed into icicles.

The student will bear in mind that in ascending mountains in the torrid zone, we enter upon regions of perpetual snow at a height of 16,000 or 18,000 feet. The snow line—that is, where snow is found throughout the year—is reached at a less and less height, as we recede towards the poles; and, in the polar regions, snow is constant at ordinary levels.

Other circumstances being favorable, we may find glaciers in those valleys which lead up into the regions of perpetual snow.

The glaciers of the Swiss Alps have been studied with great care by De Saussure, Venetz, Charpentier, Agassiz and Guyot, and Forbes.

The writings of these eminent men upon this subject are full of interest and instruction.

Agassiz and Guyot devoted eight or ten years to the study of the Swiss Glaciers. While the former gave his attention more especially to their present condition, carrying on the most extensive and accurate surveys, particularly upon the Glacier of the Aar, which resulted in making known to the world the laws which govern glacier motion,—the latter spent seven years in tracing out, for the first time, the position of the boulders, and marking the extent of the glaciers in past times.

All the high valleys of the Alps contain glaciers which vary, according to local circumstances, in length, breadth, and thickness, and in respect to the height at which they terminate. In some cases they are only a few miles long; in others, fifteen or twenty miles. They vary in width from 500 or 600 feet to two or three miles. In thickness, they vary from 100 to 200 feet; in some cases, however, they are 500 or 600, and even 1000 feet thick. They terminate at a height of 4000 to 6000 or 7000 feet, and sometimes descend as low as 3000 feet above the level of the sea.

A glacier descends towards the plain till heat melts it away. The mass being so enormous, the glacier often extends into the regions of cultivation, and sometimes destroys the labors of the husbandman.

The rate of movement differs in different glaciers, and in different parts of the same glacier. It may be stated, in general, that glaciers move down the valley from 3 or 4 inches to 3 feet a day; the former being the more common rate.

There are two prominent theories about the cause of glacier motion. One theory regards the glacier as a plastic body whose particles move over one another like those of fluids; and hence the mass moves onward by the force of gravity. The other theory finds the main cause of motion in the freezing and consequent expansion of water in the capillary tubes and fissures which permeate the whole mass. This expansion at innumerable points causes the whole glacier to dilate, thus producing an onward movement.

The geological changes which glaciers produce will be readily understood. Encased as they are in the mountain valleys, they receive the rocks and earth that are detached by the frosts from the steep flanks and peaks of the adjacent mountains. The result is, that all along the glacier, on both sides, we find an accumulation of loose material. These accumulations are called *Lateral Moraines*.

When two glaciers unite, the lateral moraines on each outer side continue on as before, but the lateral moraines on each inner side unite in one ridge along the middle of the newly-formed glacier. This accumulation is called the *Medial Moraine*, and always indicates a compound glacier.

The accompanying cut gives the student a good

Fig. 182.

Upper part of the Glacier of Viesch.

232 AQUEOUS AGENCIES.

idea of a glacier with both lateral and medial moraines.

The glacier pushes before it the loose materials which it meets in its course, and this accumulation is called the *Terminal Moraine*. The accompanying cut gives a good idea of the termination of a glacier, and its terminal moraine.

Fig. 183.

Lower part of the Glacier of Viesch.

A stream of water, as is indicated in the cut, always issues from beneath the lower end of the glacier. Most of the rivers of Central Europe have their origin beneath these ice masses.

The medial moraine is borne onward without essential alteration of its materials; but the rocks of the lateral moraines are more or less rounded, smoothed, and often polished against the rocky sides of the valley, which are also furrowed and smoothed by the same process. Rocks from the lateral moraines often get between the glaciers and the walls of the valley, and are there rolled over and rounded; or getting beneath the glacier, are ground to sand and mud. The lower surface being thickly set with pebbles and fragments of rock, the glacier acts like a mighty rasp, as it drags its icy bulk along its course.

As the circumstances which produce glaciers are constant, or nearly so, the result is an unceasing transport of the rocks and other loose materials of the higher regions to lower levels; and as the glacier is continually melting away at its lower extremity, the moraines are dropped in the same relative positions which they held when connected with the glacier itself.

The position of the lower extremity of a glacier varies from year to year. In cold summers, the glacier advances farther towards the plain; and in a hot summer, this extension is melted away. Then the effects it has produced can be observed. Besides the moraines which it leaves, the rocks on the bottom and sides of the valley, which have been grooved and polished by it, are open to view. Thus we learn the effects that glaciers are now producing.

234 AQUEOUS AGENCIES.

Fig. 184.

Rock striated by the action of a Glacier, Alps.

The moraines, the furrowed and polished rocks which we find above, and beyond the present limits of glaciers, show their extent in past times. The sides of the valleys of the Alps, from 1000 to 2500 feet above the present surface of the glacier, are scratched and furrowed, and moraines are found at that great height.

The evidence is conclusive that the Great Valley of Switzerland, between the Alps and the Jura Mountains, was once filled with glaciers, which transported the boulders that now lie scattered on the sides and summits of the Jura. These mountains are limestone; but their sides, towards the Alps, are covered with blocks of granite identical with that of Mont Blanc and vicinity, whence there is no doubt they were transported. One of these blocks, called Pierre à Bot, resting on the Jura, 800 feet above Lake Neufchatel, contains 40,000 cubic feet.

ICEBERGS. In the polar regions glaciers descend to the sea, and, becoming detached from the land, drift away as icebergs. Icebergs also form by the freezing of the water near the shore, and by additions of snow blown from the land. Situated often near high cliffs, they constantly receive blocks and fragments of stone, and are thus filled from the bottom to the top with loose material. At length the whole is undermined by the waves, and drifts away.

Icebergs are often of enormous size, fully justifying the appellation of iceberg or ice-mountain. Scoresby saw five hundred in latitude 70° north; some of the largest were a mile in circumference, and 100 or 200 feet above the surface of the water, and loaded with earth and rocks. Some of these bergs, according to his estimate, contained 50,000 to 100,000 tons of loose material.

As ice floats with eight-ninths of its bulk below the surface, the true thickness of the berg is nine times its height above the surface of the water.

In Baffin's Bay, Sir John Ross found many icebergs stranded in 1500 feet of water. In 1839, an iceberg about 300 feet high was seen in latitude 61° south, and 1400 miles from any known land. This contained a large mass of rock. The British steamer Acadia observed one in May, 1842, which was 400 or 500 feet high.

At some seasons of the year, scarcely a voyage is made between America and Europe without meeting with these floating ice-mountains.

Icebergs drift towards the equatorial regions, lowering the temperature of the waters through which, and that of the countries near which, they pass. They scatter along their course the earth and fragments of rock which are continually falling out of the melting mass; they plough along when they strike a bottom of loose material, crush submarine hills of slate rocks, and break off and furrow the summits of submarine mountains.

We find in the study of glaciers and icebergs, an explanation, more or less complete, of the Drift phenomena, described in the 8th Section of the 8th Chapter.

The results which glaciers are now producing,— such as transporting rocks and other loose materials, furrowing and polishing the rocks over which they pass, and piling up moraines—are analogous to, and mostly identical with the phenomena of the Drift. But whether the transported boulders, and the furrowed and polished rocks of North America, are the result of ice moving over the surface as the glaciers now move in the valleys of the Alps, or whether these results were produced by icebergs drifted over the continent when submerged, is still a question upon which geologists are divided.

It may be regarded, however, as perfectly established, that ice, in some form, has done the work. Mere currents of water could never have transported boulders across deep valleys, nor could they have produced the perfectly straight and parallel grooves upon the rocks in place. The action of running water can always be readily distinguished from glacial action, because the former has a sinuous course, while the latter moves in straight lines.

As there is no longer a doubt that the Drift phenomena were produced by ice, it only remains to be decided whether the ice moved over the surface of the country while it was above the water, or drifted in the form of icebergs over the country submerged. This question can never be settled by speculation, but only by the most careful deductions from facts, some of which, there is reason to believe, are yet to be collected.

It should be borne in mind that the agency which furrowed and polished the rocks of this country, accommodated itself to the highest elevations, within the limits before specified, and to the deepest depressions, and moved over great distances with an undeviating course. Now the bottom of a floating iceberg strikes only those points that lie in or near a given plane. That is, if a berg strike the high summits, it does not touch the deep valleys; and if it moves through the deep valleys, it does not strike the summits.

Again, icebergs are more or less influenced in their course by the winds; and instead of moving straight onward when they strike a submarine ridge, they turn around in the direction of the greatest freedom of motion

The above facts seem to indicate that floating icebergs were not the chief agents which furrowed and polished the rocks of North America. On the contrary, the facts seem to favor the theory, that the Drift phenomena were mainly produced by ice, somewhat analogous to glaciers, moving from north to south over the whole Drift region.

Without doubt the continent has been submerged since that great ice period; and during its gradual elevation, waves, currents, and icebergs have contributed largely to the appearance which the Drift regions now present.

It will be seen from our brief examination of aqueous agencies, that they all tend to degrade the continents, and fill up the sea; and if there were no other agencies at work, those already specified would, in the course of time, reduce the highest lands to a level with the ocean; for if any part of such a work be accomplished in one year, time only would be requisite to accomplish the whole. But there are other agencies in operation, whose results are often opposite to those specified as produced by water, under its different forms and conditions. These will be spoken of in the next section.

SECTION II.

IGNEOUS AGENCIES.

Igneous agencies, and their results, may be treated of under the following heads,—Volcanoes, Earthquakes, and Thermal Springs.

A VOLCANO is a mountain from whose sides or summit, either lava, cinders, stones, and hot vapors, or all of these, have been ejected. The opening whence these are thrown out is called a crater.

In America volcanoes occur on a grand scale, along the western coast, from Southern Chili to the 20th parallel in Mexico. Volcanoes also occur in California and Oregon. In the West Indies, a line of volcanoes extends from the island of St. Vincent to the island of Guadaloupe.

A line of volcanoes extends from Alaska along the Aleutian Islands to Kamtschatka, thence south, through all the islands along the eastern coast of Asia, to the East Indies, where they reach their most wonderful development, about forty occurring in Java.

A line of volcanoes extends from the eastern part of New Guinea, through the New Hebrides and Friendly Islands, to Central New Zealand. Southern

Europe and the adjacent islands constitute another volcanic region. Iceland is noted as a region of volcanoes. The Azores the Cape Verde, and the Sandwich Islands, are volcanic. The last-named islands have furnished some of the most interesting examples of eruptions in modern times.

Over the regions specified above, are volcanoes either constantly or occasionally active. Of such volcanoes there are 24 in Europe, 11 in Africa, 46 in Asia, 114 in America, and 108 in Oceanica. About two-thirds of these are situated upon islands.

The indications of an approaching eruption, are rumbling sounds, trembling of the earth, increase of smoke from the crater, and explosions. At length cinders and fragments of rock are ejected, and sometimes molten rock pours over the edge of the crater, and descends the mountain to the plain.

A volume might be occupied in descriptions of those eruptions of which we have authentic accounts; but there is room to notice only a few. We will select from those that have taken place within the Christian Era.

Previous to A. D. 63, Vesuvius was regarded only as an ordinary mountain. None, except students of nature, suspected that it was a slumbering volcano, that had in past times devastated the region around, and that it might again send forth showers of ashes, stones, and molten floods. Its sides were adorned with fertile fields, and cities flourished at its base, where a numerous population engaged in all the vocations of life, participated in its joys and hopes, untroubled with one thought of the destruction then slumbering beneath the mountain. But this year the inhabitants were startled by an earthquake, and the shocks were continued till the year 79 following, when a dark cloud of vapor, pierced now and then by vivid flashes of light, appeared rising from the summit, and soon the volcano broke forth in all its awful power, laying waste the fair fields upon its flanks, and overwhelming, and burying from human view, Herculaneum, Pompeii, and Stabiæ.

Examinations have shown that these cities were not destroyed by lava, but by other volcanic products, such as sand, ashes, cinders, and fragments of rock, though Herculaneum has subsequently been repeatedly overflowed with lava.

Not only were the cities buried in this loose material, but the buildings, cellars, and vaults, were filled by currents of mud produced by the copious showers which probably ensued from the condensation of the vapors.

Thus these cities were blotted out, and forgotten for almost seventeen centuries; and had they not been discovered by excavation, their existence might have been denied, so meagre and indefinite are the accounts in history concerning them. But all questions in regard to their existence and fate have been settled, by the restoration of the cities themselves to the light of day.

In 1713, the workmen in sinking a well, were arrested in their progress, by striking upon the theatre of Herculaneum. This city was buried to the depth of about 100 feet.

Pompeii was not discovered till 1750, although covered above the houses less than 20 feet deep.

Extensive examinations have now been made in both cities. It appears that most of the inhabitants escaped, and with a considerable portion of their movable property. But some were left behind. At Pompeii, the skeletons of two victims were found in the stocks; and, in a vault, whither they had probably fled for safety, were the skeletons of seventeen more. One of these, a woman with an infant in her arms, has left her impress upon the rock, and her bones are encircled by rings and chains of gold that adorned her in life!

The streets, the houses, and the shops, are just as the inhabitants left them in their haste, almost 1700 years before;—the pavements of lava, with deep ruts worn by the ancient carriage wheels; the names of the owners over the doors of the houses; the writings on the walls of the soldiers' barracks, the frescoed paintings, as bright as though put on but yesterday; fabrics in the shops still showing their texture and vessels of fruit so well preserved as to be easily recognised; bread retaining the stamp of the baker, and medicine on the apothecary's counter, still showing its real nature.

Since the year 79, there have been many eruptions of Vesuvius, sometimes with centuries intervening. In 1631, an eruption from this mountain destroyed Resina, a town that had been built over Herculaneum. In 1794, the lava from this volcano overflowed Torre del Greco, filling the streets, and destroying more than four hundred persons. It is estimated that twenty-two million cubic yards of lava were thrown out at this eruption. The principal street of the town is now cut through this consolidated matter, and the rock which in its molten state destroyed the former houses, was quarried to build new ones.

During the eruption of 1822, the crater of Vesuvius, which had for many years been filling up, was cleared of the accumulated material, and a gulf was formed more than 1000 feet deep, and three-fourths of a mile in diameter; and 800 feet of the top of the mountain was carried away.

There was a violent eruption from this mountain in 1850, and eruptions of greater or less extent have taken place almost every year since.

Etna, on the island of Sicily, has been more or less active from the earliest times of which we have any account. This mountain is 11,000 feet high, and

90 miles in circumference. The lower portions are covered with fertile fields, and are thickly settled; the higher portions are covered with forests of oak, chestnut, and pine; and the highest regions are barren lava, and other volcanic products. More than eighty minor volcanic cones are scattered over its flanks, some of them 700 feet high.

In 1669, the lava from Etna overran fourteen towns and villages, before it reached Catania, whose walls had been raised to the height of sixty feet, as a protection against the molten floods from this mountain. Here the lava collected till it rose above the high wall, and poured in a fiery flood upon the city. "The traveller may now see the solid lava curling over the top of the rampart, as if still in the very act of falling." After destroying a part of the town, it still flowed on, and in a stream 40 feet deep, and 1800 feet broad, entered the sea.

During this eruption a fissure, six feet wide, and of unknown depth, opened upon the mountain, for the distance of twelve miles. The filling of this fissure with melted lava would produce a genuine dike, such as has been described in the first part of this work, and such as now occur in great numbers, in Etna, and in other volcanoes.

In 1750–60, Jorullo, in Mexico, experienced a violent eruption. Six volcanic cones were formed in the very district where before were fields of sugar-cane and indigo. Forty years afterwards, Humboldt found the mass of matter produced at this eruption, still hot.

In 1783, Skapter Jokul, in Iceland, sent forth two streams of lava which flowed in opposite directions. One of these streams was 50 miles long and 12 broad, and the other 40 miles long and 7 broad; each averaging 100 feet deep, and when pressed into gorges, as was the case in some parts of the course, 600 feet deep. The eruption continued for two years, and destroyed twenty villages, and 9000 inhabitants.

Imagine a river of molten rock, 90 miles long, 7 to 12 miles broad, and 100 feet deep, and you have some idea of the amount of matter poured out of Skapter Jokul in the eruption which commenced in 1783.

In 1815, a violent eruption took place on the island of Sumbawa. The explosions were heard at the distance of nearly a thousand miles. The falling ashes crushed houses 40 miles distant from the place of the eruption, and they so filled the air in Java as to cause total darkness in the daytime. Floating cinders so covered the sea west of Sumatra, that ships made their way through them with difficulty.

The lava flowed over the land, and entered the sea. Whirlwinds swept over the island, tearing up trees, and bearing off men, horses, and cattle. Of 12,000 inhabitants, only twenty-six survived the awful catastrophe.

The most remarkable eruptions of the present century have taken place upon Hawaii, from Mauna Loa and the craters upon its sides.

Mauna Loa is 13,760 feet high. Kilauea, 9790 feet below the summit of the former, is a crater 16,000 feet long, 7500 feet wide, and 1000 feet deep. From

Fig. 185.

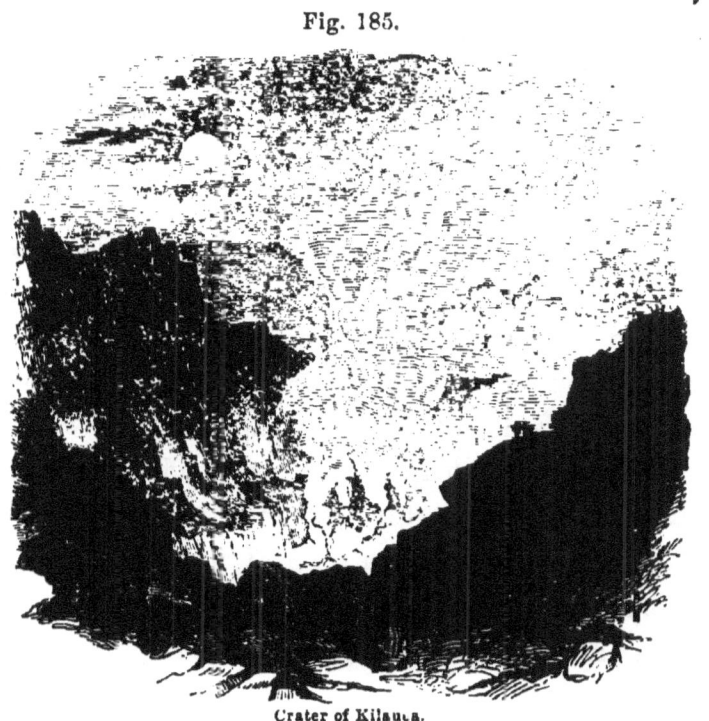

Crater of Kilauea.

this crater there was a great eruption in 1823, which overflowed extensive regions in the south-west part of Hawaii, and at last the lava entered the sea in a stream four or five miles wide. The amount of lava thrown out at this eruption is estimated at 27,000,000,000 cubic feet.

Another eruption from this crater occurred in 1832, and another still in 1840. Of the latter we have very definite accounts from Rev. Mr. Coan, whom I quote below. Since 1832 the crater had been gradually filling up until the lava was 400 or 500 feet above its ordinary level, and for some time before the eruption, this molten mass was "raging like the ocean when lashed into fury by a tempest." At length, on the 1st of June, 1840, the lava, having made its way through subterranean fissures, began its flow through outlets several miles below the true crater, and "sweeping forest, hamlet, plantation, and everything before it, rolled down with resistless energy to the sea, where, leaping a precipice of forty or fifty feet, it poured in one vast cataract of fire into the deep below, with loud detonations, fearful hissings, and a thousand unearthly and indescribable sounds.

"Imagine to yourself a river of fused minerals, of the breadth and depth of Niagara, and of a gory red, falling in one emblazoned sheet, one raging torrent, into the ocean!

"The atmosphere in all directions was filled with ashes, spray, and gases; while the burning lava, as it fell into the water, was shivered into millions of minute particles, and, being thrown back into the air, fell in showers of sand on all the surrounding country.

"The coast was extended into the sea a quarter of a mile. Three hills of scoria and sand were also formed in the sea, the lowest about two hundred, and the highest about three hundred feet high.

"For three weeks this terrific river disgorged itself into the sea, with little abatement. The waters were heated for twenty miles along the coast, and multitudes of fishes were killed.

"The breadth of the stream, where it fell into the sea, is about half a mile; but inland it varies from one to four or five miles in width, conforming, like a river, to the face of the country over which it flowed. The depth varies from 10 feet to 200, according to the inequalities over which it passed. The whole course of the stream, from Kilauea to the sea, is about 40 miles.

"During the flow, night was converted into day in all eastern Hawaii. The light rose and spread like the morning upon the mountains; and its glare was seen on the opposite side of the island. It was also distinctly visible for more than one hundred miles at sea; and at the distance of forty miles fine print could be read at midnight."

According to Professor J. D. Dana, 15,400,000,000 cubic feet of matter flowed from Kilauea at this eruption—a mass equal to a triangular ridge 800 feet high, two miles long, and a mile wide at base.

Mauna Loa, the great central crater, had an eruption in 1843, sending forth two streams of lava, one of which was 25 miles long, and a mile and a half wide. During this eruption, the mountain was rent for the distance of 25 miles. This mountain had another eruption in 1852, during which it sent forth a jet of liquid matter 1000 feet in diameter, and 700 feet high. Another eruption commenced in August 1855, and for a long time threatened the destruction of Hilo. For seven or eight months, a torrent of molten rocks flowed forth from the crater, forming a stream 65 or 70 miles long, and 3 to 5 miles wide.

This year, 1859, has witnessed another great eruption from Mauna Loa. It commenced suddenly on the 23d of January, and, in a single night, the molten flood ran 25 miles. Jets of liquid rock were thrown to the height of 1500 feet. Eight days after the commencement the lava reached the sea, 40 miles from the crater. Here it destroyed cocoa-nut groves, and a little village, and extended for two miles into the water.

But volcanoes are not confined to the land. We have authentic accounts of eruptions in the sea, although it is probable that only a few, among the many that have taken place, have been observed.

In 1783 a volcano broke forth in the sea off the coast of Iceland, covering the water with pumice to the distance of 150 miles. An island was formed which the Danish government called Nyöe; but it disappeared in less than a year, leaving a reef of rocks beneath the surface of the water.

In 1811, an island, called Sabrina, was formed by a volcanic eruption, near the Azores. Its cone was 300 feet above the water. It was soon levelled down by the action of the sea.

In 1831, a volcanic island appeared off the south-east coast of Sicily, where a few years before there was a depth of 600 feet of water. It is known under the name of Graham Island. It ejected volcanic products, and covered the sea with cinders. Its greatest dimensions were 200 feet in height, and 3 miles in circumference. But this also disappeared in a few months.

It will be seen by the statements already made that the force exerted in a volcanic eruption is immense. To raise lava to the summit of Mauna Loa, Dana shows that the pressure at the sea level must be 17,200 pounds to the square inch, or 2,500,000 pounds per square foot! Yet Cotopaxi, 3000 or 4000 feet higher than Mauna Loa, has ejected matter 6000 feet above its summit.

From the few examples of volcanic action here enumerated, the student may get some idea of the mighty changes volcanoes are producing.

A single eruption like that of Skapter Jokul, pours out melted rock enough to cover a township 30 miles square to the depth of 100 feet.

IGNEOUS AGENCIES. 251

Dana has shown that the Sandwich Islands, and many other islands of the Pacific, have been built up by successive volcanic eruptions; each eruption adding a new layer to some part of the accumulating mountain. Hence the stratification which volcanic rocks sometimes exhibit. Still, there is no difficulty in distinguishing volcanic productions from aqueous deposits; the former always indicating their fiery origin.

In the numerous and extended fissures, formed and filled with lava during an eruption, we have a full explanation of the dikes which we find in other places where the volcanic fires have long since ceased to act.

In the vast amount of matter thrown out by volcanoes, we find positive proof that the mountain itself does not furnish the material which thus flows out like a mighty torrent; for the amount of a single eruption is sometimes greater than the whole mass of the mountain. Hence, we are compelled to look to a vast reservoir of melted matter for this great supply. In a word, the volcano is but the outlet to the channel or channels connecting with the molten interior of our globe.

The immediate cause of an eruption is undoubtedly the force produced by the expansion of steam and gases in contact with the molten mass beneath the mountain.

In closing the remarks on volcanoes a word may be added about lava itself.

Lava varies greatly in composition, structure, and color. In some cases it is very compact; in others, so porous and sponge-like that it will readily float upon the water. The varieties in structure result mainly from the different circumstances under which the molten matter cools. When cooled under the pressure of a superincumbent mass, it generally forms compact lava. The porous varieties are called scoria and pumice, and result from the cooling of matter while expanded with the contained vapors and gases. In fact, scoria and pumice are often but the froth or foam of the volcano.

As the lava is thrown into the air, the wind often spins it into fine threads of great length. Great quantities of this are found at Kilauea, where it is called "Pele's hair."

THERMAL SPRINGS are intimately connected with volcanic phenomena, and deserve notice in this place. They occur in almost every country, and remote from volcanoes, as well as near them. There are thermal springs in Virginia, Arkansas, Oregon, the Utah Basin, and in other parts of North America.

The Geysers, in the south-western part of Iceland, have long been noted. The Great Geyser issues from an elevated basin, fifty feet in diameter, which gradually contracts into a pipe or tube, eight or ten feet in diameter, with a perpendicular depth of about eighty feet. The basin is sometimes empty, but is generally filled with boiling water.

Sometimes a column of water is thrown up 100 or 200 feet. This violent action lasts only a few minutes. After the water is thrown out of the pipe, an immense quantity of steam rushes up with a deafening roar; then all is quiet for a time. The principal changes produced by thermal springs is the deposition of silicious matter.

EARTHQUAKES. Some of the phenomena of earthquakes are rumbling sounds, undulatory movements of the ground, the opening of vast fissures in the earth, the drying up of fountains, the elevation and depression of whole districts or countries. A few examples, selected from the many which are recorded, will serve our present purpose.

In November, 1755, an earthquake occurred at Lisbon, which in six minutes destroyed a large part of the city, and 60,000 inhabitants. The loftiest mountains of Portugal were rent, and huge fragments were thrown down into the valleys. Near Lisbon, the waters of the ocean rolled back, then rose again fifty feet above their ordinary level. The land in all the adjacent countries was convulsed, and in the north of Africa, near Morocco, a town with 8000 inhabitants was swallowed up.

Humboldt states that this earthquake was felt over an extent greater than four times the area of Europe.

In 1783, a great earthquake occurred in Calabria, entirely changing the face of the country, and destroying 100,000 inhabitants. The centre of greatest motion was near the town of Oppido. Vast fissures opened and closed, swallowing up the inhabitants, cattle, trees, and houses. Many of the fissures remained open after the earthquake was over.

In 1811–12, an earthquake at New Madrid, Missouri, caused the sinking of the country over an extent of 75 miles by 30, to the depth of several feet. Lakes, twenty miles in extent, were formed in the course of an hour. The earth rose in great undulations; fissures opened, and vast columns of water, sand, and other material, were thrown out to great heights.

Near New Madrid, the ground rose so as to cause a temporary reflow of the waters of the Mississippi. In some cases, the inhabitants felled trees over the opening fissures, and getting upon them escaped being swallowed up. The evidences of this earthquake are still visible.

In 1812, the city of Caraccas was reduced to ruins in a moment, by an earthquake.

In 1822, an earthquake elevated the coast of Chili from two to four feet, over an area of 100,000 square miles.

At 9 o'clock, on the 4th of January, 1843, earthquake shocks were felt throughout most of the United States, from the Mississippi to the Atlantic coast.

On February 8, 1843, the earthquake of Guadaloupe was felt from New York to the Amazon river; but the greatest intensity was at Guadaloupe. Here the earth opened, ejecting columns of water, and, closing again, engulphed men and animals.

Earthquakes are undoubtedly caused by the action of the heated interior of the globe upon its crust or outside part; but precisely how this molten matter operates to produce the earthquake, is not fully understood.

VERTICAL MOVEMENTS WITHOUT APPARENT EARTHQUAKES.

Scandinavia has been, and is now, undergoing quiet vertical movements.

The observations in respect to these movements, have been more particularly made in Sweden. That country is gradually rising; the more rapidly as we go toward the north. In fact, the extreme southern portion is gradually sinking.

That elevation has been going on, is proved by the fact, that beds of shells, of the same species as those now inhabiting the adjacent waters, are found from 100 to 200 feet above the present water level, and barnacles are attached to the rocks at that height.

These proofs exist from Gottenburg and Uddevalla to Torneo, and even to the North Cape. Similar proofs of elevation, within a comparatively recent period, exist in Norway.

That the most southern portion of Sweden is sinking gradually, is shown by the streets of the ancient seaport towns, which are now below the water level. Another proof is the fact that the sea is 100 feet nearer a large rock at Trelleborg than it was in the time of Linnæus, as shown by the record of that naturalist.

The west coast of Greenland is gradually sinking.

Since the Christian era, parts of the coast of Italy, in the region of Naples, have undergone vertical movements, amounting to 20 or 30 feet in each direction. The record of these changes is not only left on the coast itself, but also on the columns of the temple of Jupiter Serapis, at Puzzuoli.

The remains of this temple had not been noticed till 1749, at which time three pillars were found projecting several feet above a marine deposit in which they stood. Their tops were surrounded with bushes, which had prevented their previous discovery.

The position of the columns in the marine strata, furnishes us with the proof that the land had been depressed, and again elevated.

That the sea stood around these columns for a considerable length of time, is shown not only by the depth of marine strata found surrounding them, but by the fact that the columns themselves are perforated extensively by lithodomous molluscs whose shells still remain in the cavities.

Similar records found in connection with this temple, inform us that elevations and depressions, previous to those now described, have also occurred in this region.

Large regions in the southern portion of South America have been undergoing gradual elevation.

The cause of these quiet vertical movements probably lies in the fact of the gradual cooling and consequent contraction of the molten matter within. The crust of the globe, in accommodating itself to the contracting nucleus, must of necessity rise in some places and fall in others.

SECTION III.

ORGANIC AGENCIES.

Organic agencies, though silent in their operations, are none the less surely producing important geological changes.

We have already seen that the strata in many districts are mainly composed of the remains of plants and animals, and that the beds of mineral coal are wholly of vegetable origin. Plants and animals are still contributing to the formation of strata upon our globe.

The great bulk of vegetation, as remarked in another place, is carbon. Plants receive all their food through the leaves, and through the roots; the former receiving the gaseous, and the latter the liquid nourishment. There is about one gallon of carbonic acid in twenty-five hundred gallons of pure atmospheric air. Decay, combustion, and respiration, all tend to make the amount of carbonic acid greater; and the whole atmosphere would become unfit for respiration, were it not for the fact that plants receive their carbon mainly from the air. Every tree hangs out its thousands, and often its millions of leaves. Each leaf is covered with almost an innumerable number of mouths. Through these the carbonic acid of the atmosphere is taken into the leaf, where it is decomposed, the carbon retained for the growth of the plant, and the oxygen given back to the atmosphere.

In this connection I ought to notice the fact, that every branch upon the tree, every twig upon the branch, every leaf upon the twig, and the sepals, petals, and stamens of every flower, are arranged with mathematical relations to another; which not only contributes to the growth of the plant, but reveals that perfection which is found everywhere in Nature. On most plants the leaves, and consequently the branches, are arranged so that a thread drawn from one to the next above it, and from this to a third, and so on, will pass spirally around the stem; and a little careful study will show that leaves upon plants are either $\frac{1}{2}$, $\frac{1}{3}$, $\frac{2}{5}$, $\frac{3}{8}$, $\frac{5}{13}$, $\frac{8}{21}$, $\frac{13}{34}$, or $\frac{21}{55}$, of the circumference of the stem apart,—disregarding only the vertical distance which separates them. This secures the greatest possible space to each leaf, whose function it is to secure the poisonous gas of the atmosphere, and extract from it the material with which to form the woody fibre.

The forests and the little plants, then, are everywhere gathering in the carbon from the atmosphere, which they sooner or later add to the ground upon which they stand. Plants are forming peat in all the swamps of the cool countries. It is rarely formed in the hot regions, on account of the rapidity with which vegetation decays there. The amount of carbon in the peat bogs of the United States and Europe is immense, and shows us what changes are going on through the silent growth of plants.

At certain seasons of the year, drift-wood covers our large rivers that flow through timbered regions. This collects in rafts covering the river for long distances, or it becomes water-logged and sinks, and is covered by the sands. Large accumulations of vegetation are found along the Mississippi and its tributaries. Near the mouth of that river, sand, gravel, and vegetable matter, in alternate layers, have accumulated to a great depth.

In thirty-five years, a raft of drift-wood, 10 miles long, 600 feet wide, accumulated on the Atchafalaya. Over this a soil formed, giving support to a luxuriant growth of plants. A raft on the Washita concealed the surface of the river for 50 miles. Similar cases, though less extensive, are common.

The various vegetable accumulations described above, may form the coal-beds of a coming age.

Animals are constantly producing changes; and it is a remarkable fact that those which are producing the greatest changes are among the smallest in the Animal Kingdom. Silicious and calcareous deposits are constantly forming from the shields of infusorial animals. The marl which is found beneath peat in peat bogs, is wholly composed of the shields or skeletons of Infusoria. Such deposits are found at the bottoms of many of the ponds in New England. Calcareous marls are also forming in many parts of our country. In Williamstown, Vt., the bottom of a pond or lake, which was drained some years ago, is covered with calcareous marl from two to twenty feet deep.

Coral polyps especially claim our attention. These have already been described in our remarks on the Animal Kingdom. It only remains to give here some idea of the extent of the changes coral animals are producing.

Wherever circumstances favor its growth, accumulations of coral border all the islands and shores within the tropics. These accumulations are called Coral Reefs.

When a reef stands close to the shore, it is called a *Fringing Reef;* and when at a distance from the shore, a *Barrier Reef*. Both fringing and barrier reefs are often found around the same island.

Coral reefs vary in width, from a few hundred feet to a mile. The barrier reef sometimes stands 10 or 15 miles from the shore, leaving a wide channel, and often fine harbors, between itself and the land. Sometimes the barrier reef nearly encircles the island, leaving only here and there an opening; in other cases, it appears only on one or two sides.

A barrier reef sometimes surrounds several, and not unfrequently very many islands. Between a barrier reef and the island, the water is generally shallow; but, at a little distance outside the reef, it is often unfathomable.

A Coral Island or Atoll is simply a barrier reef

Fig. 186.

Coral Island with a Lagoon, South Pacific.

surrounding a body of water or lagoon, as represented in Fig. 186. Sometimes, however, the lagoon is wanting.

Coral islands seldom rise more than 8 or 10 feet above the water, being so low that the waves often dash over into the lagoon.

Both in the lagoons, and in the channels between the barrier reefs and the island, corals grow in great perfection.

It was formerly supposed that coral reefs and coral islands were built up from the great depths of the sea; but it is now known that most reef-producing polyps do not flourish in water that is more than 20 or 30 fathoms in depth. It was also formerly supposed that the annular form of coral islands was due to their having been built on the edges of submarine craters; but this is no longer believed.

The explanation of coral reefs and coral islands is one and the same. They grow on the flanks of mountains whose summits are high islands. Such islands are numerous in the Pacific; and it is well known that many of them are gradually sinking.

The upward growth of coral thus tends to counterbalance the loss of land occasioned by subsidence in those regions.

Through the influence of coral polyps, there are now two hundred islands in the Pacific, where, but for the growth of corals, there would be less than twenty.

The following ideal sections of a gradually sinking island illustrate the growth of coral reefs and the formation of Atolls.

Fig. 187 represents an island with its shores fringed with coral, constituting a Fringing Reef.

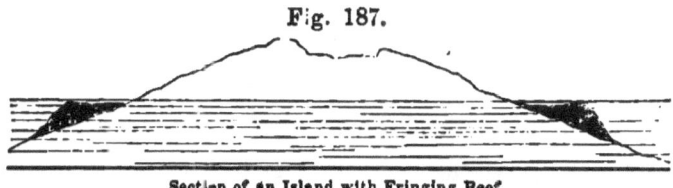

Section of an Island with Fringing Reef.

Fig. 188 shows the same island after further subsidence, and a great increase of the reef, which has now become a Barrier-Reef.

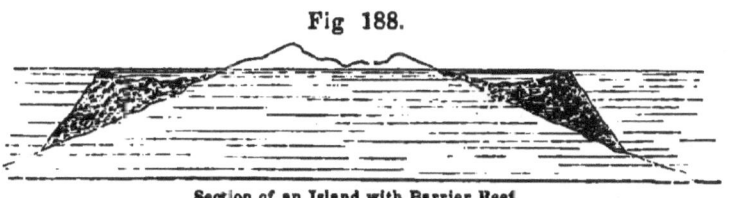

Section of an Island with Barrier Reef.

At length the island is completely submerged; the Barrier Reef becomes a Coral Island, or Atoll.

Section of a Coral Island or Atoll.

Other things being favorable, reef-growing polyps flourish wherever the winter temperature is not below 66° F.

The greatest reef region in the world extends from New Caledonia to the northeast coast of Australia. A reef 400 miles long skirts the western shore of the first-named island; and a reef extends along the north-east coast of Australia from East Cape to Torres Strait, a distance of 1000 miles. All the islands between New Caledonia and Australia are of coral growth.

The Feejee, Tonga, Navigator, Society, and Gambier Islands abound with coral reefs.

The Pamotus embrace eighty coral islands, most of them with lagoons. The Hapai group is composed wholly of coral islands. The islands of Central Archipelago, north of the Feejees, are all of coral formation, as are nearly all of the Caroline Islands.

The Sooloo Sea is a coral reef region; and coral abounds in many other parts of the East Indies. The islands of the Indian Ocean are of coral growth, or bordered by coral reefs. The Red Sea and Persian Gulf are coral regions.

Florida, south of St. Augustine, is wholly of coral formation; and extensive reefs are still growing on the coast of that peninsula. The Bermudas are of coral growth; and coral abounds in the West Indies.

Some of the coral reefs of the Pacific are from 1000 to 2000 feet thick,—so great has been the subsidence, and yet so slow, that the growing reef has kept pace with the sinking land.

Did not subsidence take place, no reef could become thicker than 20 or 30 fathoms—the maximum depth at which coral polyps flourish.

The rate at which coral reefs grow has not been satisfactorily determined; but probably it is not more than one foot in a century. According to recent investigations of Agassiz, in Florida, the rate does not exceed half a foot a century. Taking the highest of these estimates, we see what an amount of time is indicated by a reef 1000 or 2000 feet thick.

From these facts about coral polyps, we get some idea of the extent of the geological changes these little animals are producing. We also learn how the limestones of our present continents were formed; for it is settled by the most careful investigations that they are all of animal origin, and that the coral polyps have been among the most prominent agents in their production.

The common reef corals are composed mainly of carbonate of lime, 90 to 96 parts in 100 being of this substance. The remaining parts are organic matter, phosphates, fluorids, magnesia, silica, and oxyd of iron.

Therefore we find in the coral reef the materials necessary to form the limestones, and the minerals which accompany them, such as fluor spar, apatite, chondrodite, &c.

Though coral polyps have undoubtedly been the

most prominent agents in the production of the limestones, all other animals which have calcareous skeletons or shields have contributed to the same result. The shells of molluscs and other testaceous animals are contributing largely to the formation of strata in many parts of the world. It is common to find immense accumulations composed wholly of cemented shells, constituting what is called shell limestone.

Man is producing geological changes. He aids in the distribution of animals from their original centre, and causes the extinction of species. He changes the courses of rivers, rescues land from the sea; and the products of his industry go to make up a part of the strata now forming—Nature's record of the present age.

CHAPTER X.

CONCLUDING REMARKS.

WE are now prepared to take a somewhat more enlarged view of our subject, and to understand better the true relations and significance of those facts which have been pointed out in the previous pages. I propose, therefore, to devote this chapter to a statement of some of those great truths and principles which naturally flow from the facts with which the student has now been made acquainted; at the same time recalling the facts themselves, so far as necessary to bring the subject clearly before the mind.

The high antiquity of the earth is one of the great truths which geological investigations reveal.

This truth is forced upon the mind by the careful study of a single geological formation; but when we connect all the great formations and systems in their chronological order, the human mind is unable to comprehend the vast duration which opens before it.

Until a comparatively recent period, the opinion was popular that the earth is only about 6000 years old; but the records in the rocks compel us to believe that this is no more than the measure of time since the present order of things commenced.

We have looked upon the Pyramids, and the buried cities of the East, as monuments of great antiquity; and compared with the life of man, they may be so considered; but the rocks in the foundations of Nineveh, and the blocks which compose the Pyramids, were filled and covered all over with hieroglyphs when taken from the quarry; and they tell us of ages a thousand times more remote from the builders of those massive structures, than is the amount of time which separates us from them.

We shall never be able to number the centuries our planet has been in existence; we must measure its history by epochs, and not by years. But we shall gain a truer idea of the age of the earth, if we trace back the stream of time through the several geological periods, beginning with the most recent, and going back to the oldest. Let us endeavor, briefly, to do this.

By an examination of the deltas, and other superficial deposits which are comprised under Alluvium, as limited in our classification, we are carried back to a period long anterior to the creation of man.

The deep gorge at Niagara, and the deltas of the Mississippi, the Ganges, and the Nile, point us to changes that must have occupied tens of thousands of years; yet these have all taken place since the continents assumed essentially their present position, and are therefore among the modern changes that have passed upon the earth.

The changes indicated by the Drift have also occurred since the present great outlines were stamped upon the continents; but before those just specified, as is proved by the fact that the rivers have been cutting down their channels through the accumulated material of the Drift Period, and leaving terrace after terrace as if for way-marks in the march of ages.

We observe the boulder poised upon the hill, or mountain side. Could we but read its history aright, it would doubtless reveal a fund of information concerning our earth, greater than that now possessed by any human mind. The storms have beaten upon it for ages where it now stands, and the hand of time has broken many a fragment, and piled them at its base, as if to number the centuries of its age.

But the boulder has not always rested in its present position. By examination we find it entirely different from the rock upon which it rests, and is therefore not a part of that. It has journeyed far. No human eye beheld its wanderings, or saw it, at the close of its journey, rest upon the mountain. Yet its history is written. From the time it was broken from the parent ledge, till it stopped in its present place, it wrote its own story, which can be read upon the rocks to-day. There is its path in the solid rock. The rains have beaten upon that path for long years, but have not washed it away. Old Ocean has rolled his waves over it, but there it is. See what heights this boulder has travelled over—what valleys it has crossed! The Glacier, or the Iceberg, has borne it over all these, and lifted it high above the valley and poised it upon the mountain side.

Who shall number the centuries that have rolled away since it paused there in its course, or measure the time of its journey? Nay, who shall tell the ages that it occupied its place in the parent ledge, before the glacier or the iceberg wrenched it from the mass, and bore it away?

But look closer. This boulder itself, perchance is made up of pebbles, each of which was once a part of a solid rock, whose fragments, worn and polished by the lapse of centuries, are again cemented to form the rocky strata, which, after being undisturbed for ages, are broken up, and the masses borne away by the ice, and scattered upon distant mountains, where they have been so long, that Time has written *antiquity* on their decaying fronts.

Such are some of the evidences of antiquity, which the phenomena of the Drift unfold to us.

But, before the Drift, were produced all those deposits, embracing several extensive formations, which constitute the Tertiary System. It was during the Tertiary period that the strata were formed which fill the basins of London, Paris, and Vienna, to a depth several times greater than that which has accumulated since the creation of man. During this period, large tracts of America, along the Atlantic coast, were beneath the waters; and Europe was so penetrated and covered by the ocean that it presented a widely different aspect from what it presents to-day.

Yet the Tertiary period was one of life. Fishes swam those ancient waters, and molluscs crept along the bottoms. In the shallow coves where the waters of the land mingled with those of the ocean, reptiles, both large and small, basked in the sunshine, and dragged their bodies among the reeds and rushes, as they do now.

The rivers now and then are flooded, and sweep onward to the sea the foliage and trunks of palms, and cone-bearing trees, land and river shells, and the bodies of Anoplotheria, and Paleotheria, and other animals that cannot escape the descending flood; and reptiles too, that make their haunts near the rivers' mouths, are borne out, and mingled with the remains of the life of the sea.

Thus the strata accumulate, and the continents at length are lifted higher above the waters, and assume more nearly their present outlines,—the Tertiary Period is ended, yet its records of life, and death, and change remain.

But the Cretaceous System points us to a period, as

long probably as the Tertiary, which closed ere the latter began.

And before the Cretaceous, was the Oölitic Period, during which reptilian life reached its culmination. In this era, no true placental mammals lived. For thousands of years, the Plesiosaurians, Ichthyosaurians, and Pterodactyls, could well be styled the lords of creation.

Before the Oölitic, came the New Red Sandstone Period, at the commencement of which reptilian life had scarcely begun on our planet. In this period, the sandstone of Connecticut River Valley was formed, upon successive layers of which are impressed the foot-prints of animals so far removed from our time, that the naturalist hesitates, in many cases, whether to refer them to Bird or Reptile.

But the Carboniferous Period was before the New Red Sandstone,—and how shall we adequately express its great duration?

The united thickness of the coal-beds in some regions is from 100 to 200 feet, and these beds are distributed throughout ten to twenty times the thickness of strata which have accumulated since the creation of man.

When we consider the thickness of the coal-beds, and the thickness of the strata which contain them, and are interstratified with them, and when we remember that this carbon has been taken from the atmosphere by vegetation, and that the containing rocks have accumulated through the agency of water, and indicate successive vertical movements during their formation, which also implies time,—when these facts are properly collated, and their true import understood, we have an argument for antiquity which is irresistible.

Yet, the Carboniferous, embracing such a great thickness of strata, implying an immensity of time of which we can scarcely form any conception, is but a single one of those grand periods which measure off the mighty cycles of the past.

But before this period began, the Old Red Sandstone Period was past, leaving, as a faithful record of the time of its duration, stratified rocks thousands of feet in thickness, and replete with the remains of life.

But ere any, and all these named above, came the long Silurian. In this period life dawned upon our planet—marine life, yet it foreshadowed that of to-day.

Who can follow the centuries, and faithfully register their number, while the twenty or thirty thousand feet of Silurian rocks accumulated on the bottom of those ancient oceans? So slowly did the work of deposition go on in many places, that the bottom of the ocean was elevated more rapidly by the additions from the remains of animal life, than from the sediment thrown down by the waters. So stands the record to-day. Strata, ten, twenty, fifty, or a hundred or more feet in thickness, composed almost wholly of animal remains, prove this statement beyond the least shadow of a doubt.

For a period beyond the grasp of human conception, the Silurian ocean covered the greater part of the surface of the earth. Then were only scattered islands, where are now broad continents. The range of animal species was wide; the same living in various parts of the world. This period ended, but its stupendous monuments still remain, upon which we may read the story of its great duration.

Thus we have threaded our way back through the life periods of our globe, but we have not reached the source of the "Stream of Time."

Before any and all these, the solid rocks tell us of time probably as extended as the aggregate of those periods through which we have in imagination now passed, during which our earth was an uninhabited cheerless world. Nay, what ages must have rolled away, during which the earth's crust became cooled, and the waters gathered upon it; which took place before the stratified rocks even began to form.

Now, when we connect all these periods, what an ocean of time rolls up from the mighty past! We are then ready to exclaim with the Psalmist, "OF OLD HAST THOU LAID THE FOUNDATIONS OF THE EARTH."

This truth, the great antiquity of the earth, so plainly taught in nature's own records, is one which has caused the science of Geology to be looked upon with suspicion, by those who believe the Sacred Scriptures limit the age of the world to 6000 years. And this truth is often assailed as though it were the peculiar property of the geologist, while it is a truth which belongs to all mankind. It had not its origin with man, but with God. It was registered in enduring characters ere man was created. And I will not attempt to say whether it be worse to deny the truths which He has revealed in His Word, or those which He has revealed in His works.

As truly as the earth exists, so true is it that the rocks are filled with different races of animals that have successively inhabited the surface of the earth. We may argue as much as we please to show that the "plastic forces of nature" have moulded all these plants and animals in the rocks, or to show that they are not real, or that they were created in the rocks where we now find them; but let us remember one thing—that argument can never controvert facts. Galileo was compelled by arguments of force to recant his theory of the universe, but the facts remained the same; and each rising sun reiterates the same truth, which some of his time would fain have argued out of existence.

It matters not what our mere opinions are about the records in the rocks, so far as regards their truthfulness,—miles of strata filled from the bottom to the top with animals of different races, that crawled on the bottom of ancient oceans, or darted through their waters, or flew above them, or basked on the shores of estuaries, or swam the rivers, or roamed over the land, or winged their way through the air,— exist as an incontrovertible fact.

If such be the testimony of the rocks, we may safely aver that it is not contradicted by the Bible; for have not the volume of Nature and the volume of Inspiration the same Great Author? The greatest scholars of the present time are fully satisfied that the Bible does not fix the age of the earth; and that the word which is translated *day* does not always mean a period of twenty-four hours, nor always even the same amount of time, as may be readily proved by referring to the different places where it is used.

We may then regard the word *day* as used in the first chapter of Genesis as representing a period of time of great length; this interpretation is demanded by the records which God has given us in the volume of Nature,—in the rocks; and there is nothing in the Bible to forbid it.

If we regard the six days mentioned in Genesis as representing successive long periods of time, the apparent difficulty disappears, and the records agree in all their essential features. How preposterous the idea that there is any conflict between the two records! Not less so is the attempt to set either aside, by denying its true interpretation.

Another great truth which geology and connected sciences clearly reveal, is, that everything in nature is built upon a plan.

This truth is the more plainly seen and felt, the more the facts of nature are studied in their natural relations. There is no such thing as an isolated fact in the physical world. Each connects itself with another, and all are necessary to a perfect whole. To study facts of nature separately is one thing; to study them in their true relations is quite another. In one case, we are learning the alphabet; in the other, we are reading the sublimest truths which that alphabet can be used to express.

So far as we study the facts of nature in their natural relations, and understand their true significance, so far do we become acquainted with the thoughts of the Author of Nature ere creation began; for nature is but a full and tangible expression of those thoughts which were matured in the Divine Mind before the foundation of the world.

The beautiful chemical combinations of elements, the mathematical exactness of crystallized mineral forms, the structure and growth of plants, the animal kingdom as it now appears on the earth, exhibit forethought,—they reveal a plan. And in the succession of animal life upon our globe, a plan is revealed, "grand in its outlines, and beautiful in its execution."

Geology shows that the earth, during those long periods that rolled away, while it was passing from a state of chaos to the delightful garden we now find it, was inhabited by races of animals exactly suited to its condition.

No fact in geology is better established than that there has been a succession of races of animal life upon the earth, each higher in rank than the preceding. We find higher orders, and in some of the branches higher classes, as we pass from the earliest to the latest life-periods of the globe.

We find that all these races have been built upon certain plans. Radiates, Molluscs, Articulates, and Vertebrates, are the four Types after which all the animals, both of the present and the past, have been built.

Geology, then, shows that there is no such thing as the development of lower species into higher. It has been a favorite notion with many people, and some philosophers, that the higher animals are only the results of continued improvements on the lowest form of life. Our science shows this notion to be entirely false. Animals of the four great Types appeared upon the earth in profusion simultaneously. Hence neither branch, nor type, could be developed from another.

Upon the great plan wrought out in the Divine Mind the successive races have appeared. Each succeeding race bears the impress of the original idea, yet it is an entirely new creation. Nature has not repeated herself, but everywhere is the greatest diversity consistent with unity.

If we look at the Branch of Vertebrates, the highest of the animal kingdom, we find it represented in the earliest ages, by the lowest class, Fishes. Ages later, Reptiles are introduced; later still, Birds are added; and in the Tertiary the true Mammals make their appearance; and, last of all, Man appears upon the earth; yet the plan was unaltered during all these changes, and remains the same to-day, as when God spoke life into existence. The great idea which found its highest expression in Man, was shadowed forth in the earliest paleozoic fishes.

How significant these facts! The race of Vertebrates did not end with fishes, nor reptiles, nor birds, nor mere mammals, but with Man. The highest expression possible upon the Vertebral plan had not been made till the introduction of him whom God created in His own image!

The great plan of the Creator is still farther revealed, when we consider the wonderful provisions He has made for the benefit of His creatures; and especially for man, His last creation; and not a plan only, but Divine Benevolence is strikingly exhibited, and to a degree excelled only by that presented in the GIFT which was made as a provision for man's immaterial and higher nature.

Who reflects upon the wonderful supply of all those materials so useful and necessary in the various pursuits of life, and is not fully convinced that the earth has been going through a long series of changes, preparatory to the reception of man?

The earth was filled with rocks, and metals, and treasures necessary for man's use, ages before he was created. The vast storehouse of fuel, in the form of coal, which is now locked up in the earth, subject to the demands of human labor, floated in the atmosphere of an age long since past. At that age, neither man nor any land animal could exist, as the air was charged with a deadly gas.

The carbonic acid might have been swept away by a single stroke of the Divine Hand, but a great plan was to be fulfilled; the earth was to become the home of man, and this carbon was to be preserved for his use. The vegetation secured the carbon; and thus the air was purified. The vegetation became entombed in the bosom of the earth, and after slumbering there for untold centuries, comes to-day to benefit and bless mankind.

This carbon was buried far beneath the rocky strata; but the rocky covering of the earth has been broken, and its treasures brought within the reach of human industry. Every page of geologic history unfolds the wonderful provision which the Author of Nature made for the advent of His last creation.

In a proper contemplation of the facts registered in the crust of the earth, what a field opens before us! We begin to see how vast are the works of God. The mind seems to run back through the long-slumbering ages, and note the changes our planet has undergone. It gets a glimpse of the scenes as they passed in the slow, yet ever revolving centuries. We get a faint view of the Great Plan the Author of Nature has made and carried out. A world of disorder is transformed into one of life and beauty; peopled at different stages of its existence with beings suited to its condition; improved with each successive change, until it is as a delightful garden, where the fragrance of flowers sweetens every breeze, and crystal streams flow, whose ascending vapors make green the overhanging branches, where sweet warblers sing from fullness of joy;—when Man is created, endowed with those god-like powers and faculties which enable him to fathom the past to some extent, to comprehend and enjoy the present, and to look forward with bright anticipations to the GREAT FUTURE.

GLOSSARY.

GLOSSARY

OF SCIENTIFIC TERMS OCCURRING IN THE TEXT OF THIS WORK.

[It will be observed that the arrangement adopted in the Glossary presents, first, after the word itself, its pronunciation; secondly, its etymology; thirdly, its signification.]

A.

ACALEPHS, *ac'-ă-leefs*—etym., αχαληφη, *ac-a-lee'-fee*, nettle—marine animals which cause a stinging sensation when touched by the hand.

ACEPHALS, *a-sef'-als*—etym., *a*, without, and κεφαλη, *kef'-a-lee*, head—headless molluscs, as the clam and oyster.

ACICULAR, *ă-sic'-u-lar*—etym., *acus* (*a'-cus*), needle—needle-shaped.

ACROGENS, *ac'-ro-jens*—etym., αχρος, *ac'-ros*, top, and γενος, *gen'-os*, growth—growing at the extremity.

ACTINIA, *ac-tin'-e-a*,—etym., ακτιν, *ac'-tin*, ray—animals having a circle of tentacles, like rays, around the mouth, as the *sea-anemone* (*ă-nem'-o-ne*).

ACTINOID, *ac'-tin-oid*—etym., ἀκτιν, *ac'-tin*, ray, and ειδος, *i'-dos*, form—like an Actinia.

ACTINOLITE, *ac-tin'-o-lite*—etym., ακτιν, *ac-tin*, ray, and λιθος, *li'-thos*, stone—green hornblende.

AGATE, *ag'-ate*—etym., γαγατης, *gă-ga'-tees*—a semi-pellucid, uncrystallized variety of quartz, of varied tints arranged in stripes or clouds. It is used for seals, cups, &c.,

and is named, according to some, from the stream in Sicily on whose banks it was found.

ALABASTER, *al-a-bas'-ter*—etym., αλαβαστρον, *al-a-bas'-tron*, a variety of gypsum.

ALBITE, *al'-bite*—etym., *albus*, white—a variety of feldspar.

ALCYONOIDS, *al'-se-o-noids*—etym., άλς, *hals*, the sea, κυω, *ku'-o*, to breed, and ειδος, form—an order of polyps.

ALGÆ, *aljee*—plural of *alga*, sea-weed—marine plants.

ALLUVIUM, *al-lū'-ve-um*—compounded of *ad*, to or on, and *luo*, to wash—sandy, earthy, or stony matter, washed away from its original place, and deposited on the surface elsewhere.

ALUMINA, *al-ū'-me-na*—etym., *al-u'-men*, clay—oxyd of aluminium.

ALUMINIUM, *al-ū-min'-e-um*—etym., as above—the metallic base of clay.

AMETHYST, *am'-e-thyst*—etym., αμεθυστος, *a-me-thūs'-tos*, not inebriating—a species of quartz, valued for jewellery, and, in ancient times, used for drinking cups, as a supposed protection against intoxication.

AMMONITE—named from its resemblance to the horns on the statue of Jupiter Ammon, resembling those of a ram —a genus of molluscous animals resembling the nautilus.

AMYGDALOID, *a-mig'-dal-oid*—etym., αμυγδαλα, *a-mūg'-da-la*, an almond, and ειδος, *i'-dos*, form—minerals imbedded in trap rock, like almonds in a cake.

ANALCIME, *an-al'-sim*—etym., αναλκις, *an-al'-kis*, weak—a zeolite, giving by friction, a weak electricity. See ZEOLITE.

ANDALUSITE, *an-da-lu'-site*, a silicious mineral, found in thick lamellar forms, or in rhombic prisms. The name is derived from Andalusia, in Spain, where it was first observed.

ANELLIDES, *an-el'-e-deez*—etym., *annellus (an-nel'-us)*, a little ring—worms whose body seems to be composed of a succession of little rings.

ANGIOSPERMS, *an'-je-o-sperms*—etym., αγγειον, *ang-gi'-on*, vessel, and σπερμα, *sper-ma*, seed—plants having seed enclosed in a pericarp. See PERICARP.

ANOPHYTES, *an'-o-fites*—etym., ανω, *a'-no*, upward, and ψυω, *fu'-o*, to grow—shooting upward, erect.

ANOPLOTHERIUM, *an-o-plo-the'-rē-um*—etym., ανοπλος, *an-op'-los*, unarmed, and θηριον, *the'-ri-on*, wild beast—a fossil extinct quadruped, resembling a pig, and deficient in claws, hoofs, horns, or other means of defence.

ANTENNÆ, *an-ten'-nee*—etym., *antenna*, a sailyard—the horns of insects, resembling somewhat, in form, the yards of an ancient ship.

ANTHRACITE—etym., ανθραξ, *an'-thrax*, coal—a species of mineral coal.

ANTIMONY, *an'-te-mo-ny*—(etym., uncertain)—an ore, consisting of sulphur and a metallic base.

APATITE, *ap'-a-tite*—etym., απαταω, *ap-a-ta'-o*, to deceive—native phosphate of lime, whose deceptive appearance has often caused it to be taken for other minerals.

APHIDES, *af'-e-deez*—etym., *aphis*, ā'-fis, plant-louse—minute insects infesting plants.

APOPHYLLITE, *ă-pof'-fil-ite*—etym., ἀπω, *a'-po*, from, and φυλλον, *ful'-lon*, leaf—a mineral occurring in pearly, laminated masses, or in square prisms, and easily broken into plates or leaves.

AQUEOUS, *a'-que-us*—etym., *aqua*, water—pertaining to or caused by the action of water.

ARACHNIDS, *ar-ac'-nids*—etym., ἀραχνη, *ar-ac'-ne*, spider—an order of insects.

ARENACEOUS, *a-re-nā'-shus*—etym., *arena* (a-ree'na), sand—sandy.

ARGILLACEOUS, *ar-jil-a'-shus*—etym., *argilla* (ar-jil'la), clay—clayey.

ARGONAUTA, *ar-go-naw'-ta*—etym., Αργω, *ar'-go*, the name

of an ancient mythological ship, and ναυτης, *naw-teez*, sailor—a cephalopod to which the name Nautilus has been incorrectly applied.

ARRAGONITE, *ar'-ra-gon-ite*, a variety of carbonate of lime, first found in Arragon, in Spain.

ARSENIC, *ar'-sen-ic*—etym., αρσενικον, *ar-sen'-e-con*—a grayish lustrous metal, whose oxyd is the well-known poisonous substance.

ASBESTUS, *as-bes'-tus*—etym., ασβεστος, *as-best-os*, inconsumable—fibrous hornblende, which resists combustion.

ASTEROID, *as'-ter-oid*—etym., αστηρ, *as'-teer*, star, and ειδος, *i'-dos*, form—a star-shaped polyp.

ASTEROPHYLLITES, *as-ter-off'-e-lites*—etym., αστηρ, *as'-teer*, star, φυλλον, *ful'-lon*, leaf—fossils resembling, in form, a star-shaped blossom.

AUGITE, *aw'-jite*—etym., αυγη, *aw-ge*, lustre—a lustrous mineral, found in volcanic rocks.

B.

BARIUM, *bă'-re-um*, the metallic base of Baryta. See BARYTA.

BARNACLE, *bar-na-cle*. See CIRRIPEDS.

BARYTA, *bă-rī'-ta*—etym., βαρυς, *ba'-ruse*, heavy—the heaviest of the earths.

BASALT, *bă-sawlt'*, a variety of trap rock.

BATRACHIANS, *bă-trā'-ke-ans*—etym., βατραχος, *bat'-ră-kos*, frog—a class of vertebrates, including the frog, &c.

BELEMNITE, *be-lem'-nite*—etym., βελεμνον, *be-lem'-non*, arrow or dart—an extinct molluscous animal, having a long conical bone, resembling somewhat the head of an arrow or dart.

BERYL, *bĕrr'-il*—etym., βηρυλλος, *be-rūl'-os*,—a mineral occurring in bluish green six-sided prisms.

BIMANA, *bi-mā'-na*—etym., *bis*, twice or double, and *manus*, hand—two-handed animals.

GLOSSARY.

BISMUTH, a metal of a yellowish white color.
BITUMEN, *bĭ-tu'-men*—etym., *bitumen*, pitch—mineral pitch.
BIVALVES, *bĭ'-valves*—etym., *bis*, twice or double, and *valva*, shell—molluscs having shells which are composed of two pieces, as the clam and oyster.
BLENDE, *blend*—etym., *blenden* (German), to dazzle—a metallic ore often found in brown shining crystals.
BORACIC, *bo-rass'ic*, containing borax, the substance used for soldering.
BORON, *bo'-ron*, the base of boracic acid. See BORACIC.
BOTRYOIDAL, *bot'-re-oid-al*—etym., βοτρυς, *bot'-ruse*, grape-cluster, and ειδυς, *ī'-dos*, form—shaped like a bunch of grapes.
BOULDER, *bol'-der*—etym., *bowl*, to roll—a mass of rock moved from its original place by a natural process.
BRACHIOPODS, *brack'-e-o-pods*—etym., βραχιον, *bra'-ke-on*, arm, and ποδα, *poda*, feet—molluscs with members serving the double office of prehension and motion.
BROMINE, *bro'-min*—etym., βρωμος, *bro'-mos*, fetid—a marine element having a fetid odor.
BRYOZOA, *brī'-o-zo-a*—etym., βρυον, *bru'-on*, a sea-moss, and ζωον, *zo'-on*, animal—molluscous animals of a low grade, named from the resemblance indicated in the term.

C.

CADMIUM, *cad'-me-um*, a metal found in carbonate of zinc.
CÆCILIANS, *se-sil'-e-ans*—etym., *cœcus*, blind—snake-like batrachians with very small eyes.
CALC—etym., *calx*, chalk, lime—carbonate of lime.
CALCAREOUS, *cal-ca'-re-ous*—etym., as above—earth or stone containing lime.
CALCIUM, *cal'-ce-um*, the metallic basis of lime. See CALC.
CAMEO, *cam'-e-o*—etym., *cammeo*, originally, *camaieu* (*că-mâ-yoo*), shield—a gem or shell cut in relief, named from its resemblance, in form, to a shield.
CAPSULE, *cap'-sule*—etym., *capsula*, *cap'-su-la*, a box—the seed vessel of a plant.

CARBON—etym., *carbo*, coal—the principal element in charcoal.

CARBONIFEROUS—etym., *carbo*, and *fero*, to bear—strata or rocks abounding in coal.

CARNELIAN, *car-ne'-li-an*—etym., *caro, carnis*, flesh—a silicious stone of a flesh-red tint, used for seals.

CARNIVORA, *car-niv'o-ra*—etym., *caro*, flesh, and *voro*, to devour—flesh-eating animals.

CELESTINE, *se-les'-teen*—etym., *cœlestis, se-les'-tiss*, heavenly, or pertaining to the sky—a mineral so named from its peculiar tint.

CENTIPEDES, *sent'-e-pedes*—etym., *centum* (*sent'um*), hundred, and *pedes, pe'-dees*, feet—insects with many feet, but of various sizes.

CEPHALOPODS, *sef'-al-o-pods*—etym., κεφαλη, *kef'-al-ee*, head, and ποδα, *pod'a*, feet—molluscous animals whose organs of motion are attached to the head.

CETACEA, *se-ta'-she-a*—etym., *cete* (*see'tee*), whale—mammals including and resembling the whale.

CHABAZITE, *chab'-a-zite*—etym., χαβαζιως (meaning uncertain) a zeolite occurring in oblique glassy crystals.

CHALCEDONY, *cal-sed'-o-ny*—etym., *Chalcedon, cal'-se-don*—an uncrystallized, translucent, lustrous species of quartz, of whitish color, named from a town opposite to Byzantium, where it was anciently found.

CHELONIAN, *ke-lo'-ne-an*—etym., χελωνη, *ke-lo'-nee*, tortoise or turtle.

CHIASTOLITE, *ki-as'-to-lite*, χιαστος, *ki-as'-ios*, decussated or crossed—a beautiful variety of andalusite, presenting, when cut, the form of a Greek cross.

CHLORATE, *clo'-rate*—etym., χλωρος, *clo'ros*, green—a compound of chloric acid and a base. Its gas is of a greenish color.

CHLORINE, *clo'rine*, chloric gas. See CHLORATE.

CHONDRODITE, *con'-dro-dite*—etym., χονδρος, *con'-dros*, a grain—a brittle mineral, found in primary limestone.

CHROMATE, *cro'-mate*—etym., χρωμα, *cro'-ma*, color—a salt formed by chloric acid and a base. See CHROME.

CHROME, or CHROMIUM, *cro'-me-um*—etym., χρωμα, *cro'-ma*, color—a metal so called from the beautiful colors which its oxyd imparts.

CHRYSOBERYL, *cris'-o-ber-il*—etym., χρυσος, *crū'-sos*, gold, and βηρυλλιον, *be-rul'-e-on*, beryl—a yellowish green gem.

CHRYSOPRASE, *kris'-o-prase*—etym., χρυσος, *crū'sos*, gold, and πρασον, *pra'-son*, leek—a variety of quartz of golden green or leek tint.

CILIA, *sil'-e-a*—etym., *cilia* (*sil'-e-a*), eye-lids or eye-lashes—the hairy filaments on vegetable surfaces, or the filaments projecting from animal membranes.

CIRRIPEDS, *sir'-re-peds*—etym., *cirrus* (*sir'-rus*), a lock of hair, and *pedes*, feet—animals of the barnacle kind, with long, slender, curling, clasping feet.

CLINOMETER, *cli-nom'-e-ter*—etym., κλινω, *cli'-no*, to lean, and μετρον, *met'-ron*, measure—an instrument for measuring the inclination of strata.

COBALT, *co'bawlt*—etym., *Cobold*, the name of a fancied demon of the mines—the metal of this name is used for giving a permanent deep blue color to glass.

COLEOPTERA, *kŏl-e-op'-ter-a*—etym., κολεος, *col'eos*, sheath, and πτερον, *terr'on*, wing—insects with sheathed or shielded wings.

CONCHOIDAL, *cong-coï'-dal*—etym., *concha*, *cong'-ca*, shell, and ειδος, *i'-dos*, form—shell-shaped.

CONGLOMERATE, *con-glom'-er-ate*—etym., *con*, together, and *glomero*, to heap—rock formed of fragments or pebbles conglomerated and cemented by other rocky matter.

CONIFER, *cōn'-ĭ-fer*—etym., *conus*, cone, and *fero*, to bear—cone-bearing, as the fir and pine.

CORAL, *corr'-al*—etym., κορη, *cor'-ee*, maiden, and άλς, *hals*, the sea—the substance secreted within the tissues of polyps, and constituting their skeletons.

CORUNDUM, *co-run'-dum*,—a Hindoo word—adamantine spar.

CORYNE, *cor'-e-ne*—etym., κορυνη, *cor'-u-ne*, club—a genus of jelly-fishes.

CRETACEOUS, *cre-ta'-she-us*—etym., *creta*, chalk—chalk-like.

CRINOIDS, *cri'-noids*—etym., κρινον, *cri'-non*, lily, and ειδος, *i'-dos*, form—polyps having, on a jointed stem, a head resembling in form that of the lily.

CRUSTACEAN, *crus-ta'-she-an*—etym., *crusta*, crust or shell—articulates with a shelly covering, as the crab, &c.

CRYOLITE, *cry'-o-lite*—etym., κρυος, *crü'-os*, cold, and λιθος, *li'-thos*, stone—a whitish or brownish mineral occurring in foliated masses, and obtained in Greenland.

CRYPTOGAMIC, *crip-to-gam'-ic*—etym., κρυπτος, *crŭp'-tos*, concealed, and γαμος, *ga'-mos*, fructification—flowerless plants whose fructification is concealed, not apparent.

CTENOPHORÆ, *ten-of'-o-ræ*—etym., κτενος, *ten'-os*, from κτεις, *tice*, comb—an order of jelly-fishes.

CURSORES, *cur-so'-reez*—etym., *curso*, to run up and down—an order of birds, including the ostrich and the like.

CYCADS, *si'-cads*—etym., κυκεω, *kü'-ke-o*, to mix, to intermingle—fossil plants intermediate between palms, ferns, and pines.

CYCLOIDS, *si'-cloids*—etym., κυκλος, *cü'-clos*, circle, and ειδος, *i'-dos*, form—fishes with roundish scales.

D.

DATHOLITE, *dath'-o-lite*, a silicious mineral, occurring in small complex glassy crystals.

DECAPODS, *deck'-a-pods*—etym., δεκα, *deck'-a*, ten, and ποδα, *pod'-a*, feet—an order of crustaceans.

DEVONIAN, *de-vo'-ne-an*, the system of rocks named from Devonshire, England, where they were first observed.

GLOSSARY. 293

Dibranchiates, *di-brank'-e-ates*—etym., δις, *dis*, double, and βραγχιον, gill—cephalopods which have two gills.

Dicotyledons, *di-cot-e-lē'-dons*—etym., δις, *dis*, double, and κοτυληδον, *cot-u-lē'-don*, seed-lobe—plants having two seed-lobes.

Dinornis, *di-nor'-nis*—etym., δεινος, *di'-nos*, terrible, huge, and ὀρνις, *or'-nis*, bird—an extinct bird of gigantic dimensions.

Dinotherium, *di-no-thē'-re-um*—etym., δεινος, *di'-nos*, terrible, and θηριον, *thē'-re-on*, wild beast—a fossil pachyderm of gigantic size, armed with enormous tusks.

Diopside, *di-op'-sid*. See Augite, of which it is a variety.

Discophoræ, *dis-cof'-o-re*—etym., δισκος, *dis'-cos*, disk or coit, and φερω, *fer'-o*, to bear—animals having the form of a disk or plate.

Dodecahedron, *dō-dec-ă-hē'-dron*—etym., δωδεκα, *do'-de-că*, twelve, and ἑδρα, *hed'-ră*, base or side—twelve-sided.

Dolomite, *dol'-o-mite*, a granular magnesian carbonate of lime, exemplified in many species of white marble. The name is derived from the eminent French geologist Dolomieu (*do-lo-myoo*).

E.

Echinoderms, *e-ki'-no-derms*—etym., εχινος, *e-ki'nos*, hedgehog, and δερμα, *der'ma*, skin—radiate animals with an integument covered with spines or with tubercles, as the sea-urchin, or the star-fish.

Echinoids, *e-ki'-noids*—etym., εχινος, *e-ki'-nos*, hedgehog, and ειδος, form—polyps with a spiny coat.

Edentata, *e-den-ta'-tă*—etym., *e*, from, and *dens*, tooth—animals destitute of front teeth.

Emerald—etym., (modern) *esmeralda*, and (ancient) *smaragdus* (meaning uncertain)—a green-colored precious stone.

25*

ENCRINITE, *en'-cre-nite*—etym., εν, in, and κρινον, *kri'-non*, lily—a fossil bearing some resemblance, in form, to a lily.

ENDOGEN, *en'-do-jen*—etym., ἐνδον, *end'on*, within or inside, and γενος, *gen'-os*, growth—growing on the inside of the stem.

ENDOGENOUS, *en-doj'-e-nus*. See ENDOGEN.

ENTOMOSTRACA, *en-to-mos'-tră-că*—etym., εντομα, *en'-to-mă*, insects, and οστραχον, *os'-tră-con*, shell—crustaceous animals related to insects.

EOCENE, *e'-o-seen*—etym., ηως, *e'-ose*, dawn, and καινος, *ki'-nos*, recent—strata whose fossils foreshow the existing state of the animal kingdom.

EPIDOTE, *ep'-e-dote*—etym., επιδιδωμι, to enlarge—a silicious mineral, occurring in lustrous rhomboidal prisms of a greenish tint. The name applies to the enlargement of the base of the primary in some of the secondary forms.

EQUISETACEÆ, *e-que-se-tā'-she-œ*—etym., *e'-quus*, horse, and *sē'ta*, hair—resembling horse-hair.

EQUISETUM, *e-que-se'-tum*—etym., *e'-quus*, horse, and *sē'ta*, hair—the rush called "horse-tail."

EXOGEN, *ex'-o-jen*—etym., εξω, *ex'-o*, without or outside, and γενος, *gen'-os*, growth—growing on the outside of the stem.

EXOGENOUS, *ex-oj'-e-nous*. See EXOGEN.

F.

FAHRENHEIT, the name of the inventor of the thermometer in common use, and applied in honor of him to that instrument.

FAUNA, *fawn'-a*—etym., *fauni*, ancient rural deities, resembling, in part, the lower animals—the animal kingdom of a given portion of the globe.

GLOSSARY.

FLAMINGO, *flam-in'-go*—etym., *flamma*, flame—a tropical bird of a fiery red color.

FELDSPAR, or FELSPAR—etym., *feld*, field, or *fels*, rock, and *spar*, rafter—the mineral which, along with quartz and mica, composes granite rock.

FLORA—etym., *flora*, the mythological goddess of flowers—the vegetable kingdom of a given portion of the globe.

FLUOHYDRIC, or HYDROFLUORIC, *flu-o-hī'-dric, hī-dro-flu-or'-ic*—etym., ὑδωρ, *hu'-dore*, water, and fluor (*flu'-or*)—an acid obtained from fluor-spar.

FLUOR-SPAR, *flu'-or-spar*—etym., *fluo*, to flow, and *spar*, rafter—fluorid of calcium, a beautiful mineral, used for ornamental vessels, &c.

FLUORID, *flū'-or-id*—etym. as above—a compound of fluorine, with a metallic base.

FLUORINE, *flū'-or-in*—etym. as above—a gaseous element.

FOLIATED, *fo'-le-a-ted*—etym., *folium*, leaf—in leaves, leaved.

FORAMINIFERA, *for-am-in-if'-er-a*—etym., *foramen* (*fŏr-ā'-men*), opening, and *fero*, to bear—shells whose chambers are united by a small opening or perforation.

FOSSIL—etym., *fossilis* (*fos'-sil-is*), that which may be dug up—remains of plants or animals, buried in earth or rock, and found by digging.

FOSSILIFEROUS, *fos-sil-if'-er-ous*—etym., *fossilis* (see above), and *fero* (*fé'-ro*), to bear—applied to rocks which contain fossils.

FUNGUS—plural, *fungi;* adjective, *fungous*—mushroom, mushroom-like.

G.

GALENA, *ga-lee'-na*—etym., γαλεω, *ga'-le-o*, to shine—lead.

GANOIDS, *gan'-oids*—etym., γανος, *gan'-os*, brightness, and ειδος, *i'-dos*, form—fishes with bright angular scales.

GARNET—etym., *granatus* (*gran-a'-tus*), granular, resembling the pomegranate—a silicious mineral, found in beautiful, variously-colored twelve-sided crystals.

GASTEROPODS, *gas'-ter-o-pods*—etym., γαστηρ, *gas'-teer*, belly, ποδα, *pod'-a*, feet—animals which creep on the lower surface of the body, as the snail.

GEODE, *jee'-ode*—etym., γεωδης, *ge'-ō-dĕs*, resembling the earth in form—a nodule of stone containing crystals in its cavity.

GEOLOGY, *je-ol'-o-jy*—etym., γη, *ge*, or γεα, *ge'-a*, earth, λογος, *log'-os*, discourse or science—science of the earth.

GLACIER, *glace'-yer*, *glā'-sher*, *glā'-ceer*, *glass'-yer*, or *glass'-yay*, but never properly *glā-ceer'*, or *glazier* (the latter a very common error)—etym., *glace* (*glass*), ice—Alpine accumulations of ice and snow.

GLUCINA, *glu-si'-na*—etym., γλυκυς, *glu'-kŭs*, sweet—oxyd of glucinum, the salts of which have a sweet taste.

GNEISS, *gnice*—German word—a rock resembling granite in its composition, but differing in being stratified.

GRALLÆ, *gral'-lee*—etym., *grallæ*, stilts—waders, long-legged fowls which wade, and pick up their prey, in shallow water.

GRANITE, *gran'-it*—etym., *granum* (*gra'-num*), grain—rock composed of quartz, feldspar, and mica, and having a granular or grainy appearance.

GYMNOSPERM, *jim'-no-sperm*—etym., γυμνος, *gŭm'-nos*, naked, and σπερμα, *sper'-ma*, seed—not having the seed enveloped.

GYPSUM, *jip'-sum*—etym., γυψος, *gŭp'-sos*, chalk—sulphate of lime.

H.

HELIOTROPE, *he'-le-o-trope*—etym., ἥλιος, the sun, and τροπη, *tro'-pee*, turning—a variety of rhomboidal quartz.

HELMINTH, *hel'-minth*—etym., Ελμινς, *hel'-mins*, worm—an intestinal worm.

HEMATITE, *hem'-a-tite*—etym., ἁιμα, *hi'-ma*, blood—a reddish mineral, specular iron ore, or the brownish hydrated oxyd of iron.

HETEROCERCAL, *het'-er-o-serc-al*—etym., ἕτερος, *het'-er-os*, different, and κερκος, *kerk-os*, tail—applied to fishes with unequal lobed tails.

HETEROPODS, *het'-er-o-pods*—etym., ἕτερος, *het'-er-os*, other, and ποδα, *pod'-a*, feet—molluscs whose foot serves as a fin.

HEULANDITE, *hū'-land-ite*, a silicious mineral, named in honor of Heuland, a European savant.

HEXAGONAL, *hex-ag'-o-nal*—etym., ἕξ, *hex*, six, and γωνια, *go'-ne-a*, angle—six-sided and six-angled.

HIPPOPOTAMUS, *hip-po-pot'-ă-mus*—etym., ἵππος, *hip'-pos*, horse, and ποταμος, *pot'-ă-mos*, river—a huge animal frequenting rivers and their margins.

HOLOTHURIOIDS, *hol-o-thu'-re-oids*—etym., ὁλοθουριον, *hol-o-thou'-ri-on*—an order of echinoderms.

HOMOCERCAL, *hō'-mo-serc'-al*—etym., ὁμος, *hom'-os*, the same, and κερκος, *kerk'-os*, tail—applied to fishes with equal lobed tails.

HOMOLOGY, *ho-mol'-o-jy*—etym., ὁμος, *hom'-os*, the same, and λογος, *log'-os*, proportion—affinity of structure.

HORNBLENDE, *horn'-blend*—etym., *horn*, horn, and *blenden*, to dazzle or shine—a mineral in the form of crystals or masses, of all colors, but more frequently black.

HYACINTH, *hi'-a-sinth*, a variety of zircon, of a reddish or hyacinthine tint.

HYDRA, *hi'-dra*—etym., ὑδωρ, *hu'-dore*, water—a genus of freshwater polyps.

HYDROGEN, *hi'-dro-jen*—etym., ὑδωρ, *hu'-dore*, water, and γενναω, *gen'-a-o*, to produce—the gaseous element which, united with oxygen, forms water.

HYDROIDS, *hi'-droids*—etym., ὑδωρ, *hu'-dore*, water, and ειδος, *i'-dos*, form—an order of Acalephs.

I.

ICHNOLOGICAL, *ic-no-loj'-ic-al*—ιχνυς, *ic'-nuse*, footstep, and λογος, *log'-os*, discourse—pertaining to the science which treats of fossil foot-prints.

ICHTHYOSAURUS, *ic-the-o-saur'-us*—etym., ιχθυς, *ic'-thuse*, fish, and σαυρα, *saw'-ra*, lizard—a gigantic marine fossil animal, intermediate between a crocodile and a fish.

IDOCRASE, *id'-o-crase*—etym., ιδεα, *id'-e-a*, form, and χρασις, *cra'-sis*, mixture—a silicious mineral, occurring in square, yellowish, or brownish prisms.

IGNEOUS, *ig'-ne-us*—etym., *ignis*, fire—caused by the action of fire.

IGUANODON, *ig-wan'-o-don*, a huge fossil saurian, resembling the iguana (*ig-wâ'-na*), of the existing race.

INDICOLITE, *in'-di-co-lite*—etym., *indicum* (*in'-dic-um*), indigo, and λιθος, *li'-thos*, stone—a variety of tourmaline, of an indigo tint.

INFUSORIA, *in-fu-zo'-re-a*—microscopic animals, inhabiting liquids.

IODINE, *i'-o-din*—etym., ιωδης, *i'-o-dees*, resembling a violet—a substance obtained from marine plants. Its vapor is of a beautiful violet color.

IOLITE, *i'-o-lite*—etym., ιον, *i-on*, violet, and λιθος, *li'-thos*, stone—a glassy-looking mineral, showing in one direction a brownish, in another, a violet tint.

INSECTIVORA, *in-sect-iv'-o-ra*—etym., *in*, into, *seco*, to cut or divide, and *voro*, to devour—animals that prey on insects.

IRIDESCENCE, *ir-e-des'-cence*—etym., *Iris*, rainbow—coloring like the rainbow.

K.

Kaoline, *ca'-o-lin*—the Chinese term for potter's clay.
Kyanite, *ki'-an-ite*—etym., χυανος, *ku'-an-os*, azure—a silicious mineral, found in thin crystals of a bluish tint.

L.

Lamella (plural, *lamellæ*)—etym., *la-mel'-la*, a thin plate or scale.
Lamellibranchiates, *lă-mel-e-brank'-e-ates*—etym., *lamella*, plate, and βραγχιον, *brank'-e-on*, gill—having the gills in lamellæ.
Laminated, *lam'-in-a-ted*—etym., *lam'-in-a*, plate—consisting of plates, or very thin layers.
Laumonite, efflorescent Zeolite, a mineral named from its discoverer.
Lepidodendron, *lep-e-do-den'-dron*—etym., λεπις, *lep'-is*, scale, and δενδρον, *den'-dron*, tree—a scaly-barked fossil plant found in coal-beds.
Lepidolite, *lep-id'-o-lite*—etym., λεπις, *lep'-is*, scale, and λιθος, *li-thos*, stone—a scaly species of mica of lilac tint.
Lias, *li'-as*—etym., *li'-as*, a provincial English word for layer.
Lichen, *li'-ken*—etym., λειχην, *li'-keen*, rock-moss, or tree-moss.
Lignite—etym., *lignum*, wood—mineral coal, of woody texture.
Limulis, *lim'-u-lus*—etym., *limus*, sideways—a genus of crustaceans including the horse-shoe crab.
Lithia, *lith'-e-a*—etym., λιθος, *li'-thos*, stone—an alkali found in petalite, &c.
Lithodomous, *le-thod'-o-mus*—etym., λιθος, *li'-thos*, stone, and δεμω, *dem'-o*, to build—applied to molluscs which form holes in solid rocks.

M.

LODESTONE, *lode'-stone*—etym., *lœd*, to lead, and "stone"—an oxyd of iron possessing polarity.

MADREPORE, *mad'-re-pore*—etym., *madré*, *mă-dray*, spotted, and *pore*—branching coral.

MAGNESIUM, *mag-ne'-zhe-um*—etym., Magnesia, the name of the city whence, in ancient times, that substance was obtained—the metallic base of magnesia.

MALACHITE, *mal'-ă-kite*—etym., μαλαχη, *mal'-a-kee*, mallows—carbonate of copper, in color like the leaf of mallows.

MAMMAL—etym., *mamma*, breast—animals which suckle their young.

MANGANESE, *man'-gan-eez*, a metallic substance.

MARSUPIALOIDS, *mar-sūp'-e-al-oids*. See MARSUPIALS.

MARSUPIALS, *mar-sūp'-e-als*—etym., μαρσυπιον, *mar-sūp'-e-on*, purse or pouch—animals having, like the kangaroo and opossum, a pouch in which they carry their young.

MASTODON, *mas'-to-don*—etym., μαστος, *mas'-tos*, nipple, and ὀδους, *od'-ous*, tooth—a huge fossil animal resembling the elephant, and named from the conical protuberances on the surface of its grinders.

MEDUSA, *me-du'-za*, an acaleph, commonly called "sea-nettle," a gelatinous radiate of circular form, resembling the figure of the ancient mythic shield embossed with the head of Medusa.

MEGALOSAURUS, *meg-a-lo-saur'-us*—etym., μεγας, *meg'-as*, great, and σαυρα, *saw'-ra*, lizard—a gigantic fossil amphibious animal, of the saurian tribe.

MEGATHERIUM, *meg-a-the'-re-um*—etym., μεγας, *meg'-as*, great, huge, and θηριον, *the'-re-on*, wild beast—a gigantic fossil animal.

MERGANSER, *mer-gans'-er*—etym., *mergo*, to dive, and *anser*, goose—a water-fowl which dives for its prey.

METAMORPHIC, *met-a-morf'-ic*—etym., μετα, *met'-a*, trans, and μορφη, *morph'-ee*, form—rocks transformed or changed by heat from their original form or character.

METEORITE, *me'-te-o-rite*—etym., μετεωρος, *met-e-o'-ros*, lofty—a metallic or earthy body falling from the atmosphere.

MICA, *mi'-că*—etym., *mico*, to shine—the thin shining mineral miscalled "isinglass."

MICACEOUS, *mi-ca'-she-us*, containing *mica*. See MICA.

MILLIPEDES, *mil'-le-pedes*—etym., *mille* (*mil'lee*) thousand, and *pedes* (*pe'deez*), feet—many-footed insects, as the wood-louse.

MIOCENE, *mi'-o-seen*—etym., μειον, *mi'-on*, less, and καινος, *ki'-nos*, recent—rocky strata, a minority of whose fossil shells are referable to living species.

MOLLUSCA, *mol-lus'-ca*—etym., *mollis*, soft—animals with soft bodies.

MOLYBDENUM, *mo-lib-de'-num*—etym., μολυβδαινα, *mo-lŭb-dī'-na*, a mass of lead—a metal.

MONOCOTYLEDONOUS, *mon-o-cot-e-le'-do-nus*—etym., μονος, *mon'-os*, single, and κοτυληδον, *cot-u-le'-don*, seed-lobe—plants having but one cotyledon, or seed-lobe.

MORAINE, *mo-rain'*, the loose material along the sides and middle of glaciers.

MOSASAURUS, *mo'-sa-saw-rus*—etym., *Mosa*, ancient Latin name of the town of Maestricht, and σαυρος, *saw'-ros*, lizard—a fossil animal named from the place where it was first observed.

MYRIOPODS, *mir'-e-o-pods*—etym., μυρια, *mu'-re-a*, myriad, and ποδα, *pod'-a*, feet—many-footed.

N.

NATATORES, *nat-a-to'-reez*—etym., *natator*, *nă-ta'-tor*, swimmer—water-fowl.

NATROLITE, *na'-tro-lite*—etym., *natron* (*na-tron*), soda, and λιθος, *li'-thos*, stone—a zeolite containing soda.

NAUTILUS, *naw'-te-lus*—etym., ναυτιλος, *naw'-te-los* (from ναυτης, *naw-teez*, sailor, and ναυς, *naws*, ship)—a cephalopodous mollusc.

NEMATOIDS, *nem'-a-toids*—etym., νημα, *ne'-ma*, thread, and ειδος, *i'-dos*, form—long, slender, thread-like intestinal worms.

NICKEL, a magnetic metal of a reddish white color.

NITROGEN, *ni'-tro-jen*—etym., νιτρον, *ni'-tron*, nitre, and γενναω, *gen'-a-o*, to produce—the principal gaseous element of atmospheric air.

NITRATE, *ni'-trate*—etym., νιτρον, *ni'-tron*, salt—a salt formed by the union of nitric acid with a base.

NODULE, *nod'-ule*—etym., *nodus*, knot, lump—a roundish mass or knob.

NUMMULITES, *num'-u-lites*—etym., *nummus*, money, and λιθος, *li'-thos*, stone—fossil molluscs, shaped like coins.

O.

OCHRE, *o'-ker*—etym., ωχρος, *o'-cros*, pale—colored clay, commonly of a yellowish tint.

OCTAHEDRON, *oc'-ta-he-dron*—etym., οκτω, *oc'-to*, eight, and ἑδρα, *hed'-ra*, base or side—eight-sided.

ONYX, *o'-nix*—etym., ὀνυξ, *o'-nukes*, nail or claw—a variety of chalcedony, named from its fancied resemblance, in tint and surface, to the nail or claw of an animal.

OÖLITE, *o'-o-lite*—etym., ωον, *o'-on*, egg, and λιθος, *li'-thos*, stone—limestone in rounded particles like the roe or eggs of a fish.

OÖLITIC, *o'-o-lit-ic*. See OÖLITE.

OPHIDIAN, *o-fid'-e-an*—etym., οφις, *off'-is*, snake—an animal of the snake order.

ORTHOCERA, *orth-oss'-e-ra*—etym., ορθυς, *orth'-os*, straight, and κερας, horn—fossil cephalopods with straight shells.

OXYD, *ox-id*—etym., οξυς, *ox'-use*, sharp, acid—the combination of a metal with oxygen.

OXYGEN, *ox'-e-jen*—etym. as above, and γενναω, *gen'-a-o*, to generate—the atmospheric element which supports life, named from its property of generating acids.

P.

PACHYDERMS, *pack'-e-derms*—etym., παχυς, *pa'-cuse*, thick, and δερμα, *der-ma*, skin—animals distinguished by the thickness of their skin, as the elephant, &c.

PALÆONTOLOGY, *pal-e-on-tol'-o-gy*—etym., παλαιος, *pal-ay'-os*, ancient, οντα, *ont'-a*, beings, and λογος, discourse or science—the science of organic remains.

PALÆOTHERIUM, *pal-e-o-the'-ri-um*—etym., παλαιος, *pal-ay'-os*, ancient, and θηριον, *the'-re-on*, wild beast—a gigantic fossil quadruped, resembling a tapir or pig.

PALÆOZOIC, *pal'-e-o-zo-ic*—etym., παλαιος, *pal-ay'-os*, ancient, and ζωον, *zo'-on*, animal—relating to fossil remains of animals of former periods of the earth's history.

PEDUNCLE, *pe-duncle'*—etym , *pedunculus* (*pe-dunc-u-lus*), foot or stem—flower-stalk.

PEDUNCULATED, having a stem.

PENTACRINUS, *pent'-a-cri-nus*—etym., πεντε, *pen'te*, five, and κρινον, *kri'-non*, lily—a fossil named from its pentagonal (five-sided) jointed stem.

PERICARP, *per'-e-carp*—etym., περι, *per'-ri*, round, and καρπος *kar'-pos*, fruit—applied to that which surrounds the fruit or seeds.

PERMIAN, *per'-me-an*, named from Perm, in Russia.

PEROXYD, *per-ox'-id*, abounding in oxygen.

PHÆNOGAMOUS, *fe-nog'-a-mous*—etym., φαινω, *fi'-no*, to shine or show, and γαμος, *ga'-mos*, fructification—flowering plants, whose fructification is apparent, not concealed.

PHOSPHATE, *fos'-fate*—etym., φως, *fōs*, light—a salt formed by a combination of phosphoric acid with a base.

PHOSPHORUS, *fos'-fo-rus*—etym. as above, and φερω, *fer'-o*, to bring—a combustible elementary substance, burning with a highly luminous appearance.

PHYLLOPOD, *fil'-lo-pod*, or PHYLLOPODA, *fil-op'-o-da*—etym, φυλλον, *fūl'-lon*, leaf, and ποδα, *pod'-a*, feet—a tribe of crustaceans whose foot has the flat form of a leaf.

PLACOIDS, *plack'-oids*—etym., πλαξ, *plax*, plate, and ειδος, *i'-dos*, form—fishes whose covering resembles enamelled plates.

PLATINUM, *plăt'-e-num*, the heaviest of metals, and of a silvery color.

PLESIOSAURUS, *ples-e-o-sau'-rus*—etym, πλησιον, *ple'-si-ŏn*, near, and σαυρα, *saw'-ra*, lizard—a fossil amphibian, resembling the saurian or lizard tribe.

PLIOCENE, *pli'-o-seen*—etym., πλειον, *pli'-on*, more, and καινος, *ki'-nos*, recent—rocks of the tertiary period, the largest part of whose fossil shells are of recent species.

PLUMBAGO, *plum-ba'-go*—etym., *plum-bum*, lead—improperly called "black lead."

POLYPS, *pol'-ips*—etym., πολυς, *pol'-use*, many, and πους, pous, foot—radiates having many feet or tentacles.

POLYTRICHIUM, *pol-e-trik'-e-um*—etym, πολυς, *pol'-use*, many, and τριχος, *tri'-kos*, hair—a genus of mosses.

PORPHYRY, *por'-fe-ry*—etym., πορφυρα, *porf'-u-ra*, purple—unstratified rock, containing crystals of feldspar or other minerals. The term was originally applied to a reddish rock, found in Egypt.

POTASSIUM, *po-tas'-se-um*—etym, *potasse*, *po-tass'*, potash—the metallic base of pure potash.

PRASE—etym., πρασον, *pra'-son*, leek—a silicious mineral, of a leek-green color.

PREHNITE, *pren'-ite*, a pale green, lustrous mineral, of singularly diversified form, named from its first importer from Africa.

PRISMATIC, having the form of a prism.

PROTOXYD, *prōt-ox'-id*—etym., πρωτος, *pro'-tos*, first, and οξυς, *ox'-use*, sharp, acid—a combination of one equivalent of oxygen with one of a base.

PROTOZOA, *pro'-to-zo-a*—etym., πρωτος, *prō'-tos*, first or elementary, and ζωον, animal or living being—a name sometimes applied to the *infusoria*.

PTERODACTYL, *terr'-o-dac-tyl*—etym., πτερον, *terr'-on*, wing, and δακτυλος, *dac'-tu-los*, finger—a fossil reptile, somewhat resembling the bat in form.

PTEROPODS, *ter'-ro-pods*—etym., πτερον, *terr'-on*, wing, and ποδα, *pod'-a*, feet—molluscs with organs of motion like wings.

PYRITES, *pe-ri'-teez*—etym., πυρ, *pūr*, fire—a metallic ore, combining sulphur and iron or other metals. The name was originally given from its emitting sparks, when struck against steel.

PYROXENE, *pir'-ox-een*—etym., πυρ, *pūr*, fire, and ζενος, *zen-os*, stranger. See AUGITE, of which it is a variety.

Q.

QUADRUMANA, *quawd-rū'-ma-na*—etym., *quatuor* (*quat'-u-or*), four, and *manus* (*ma'-nus*), hand—animals having four hands.

QUARTZ, *quawrts*, pure silex, occurring in glassy, six-sided, prismatic crystals; an essential constituent of granite.

R.

RAPTORES, *rap-to'-reez*—etym., *rapio* (*rā'-peo*), to seize—raveners, birds of prey.

RASORES, *rā-zo'-reez*—etym., *rado* (*ra-do*), to scratch—scratchers, birds which find their food by scratching.

RENIFORM, *ren'-e-form*—etym., *renes* (*re'-neez*), kidneys, and *forma*, form—shaped like the kidneys.

RHIZODONTS, *ri'-zo-donts*—etym., ριζα, *ri'-za*, root, and οδους, *od'-ous*, οδοντος, *o-dont'-os*, tooth.

RHOMBOHEDRON, *rom'-bo-he-dron*—etym., ρομβος, *rom'-bos*, rhomb, and εδρα, *he'-dra*, side—a solid contained by six equal rhombic planes.

RODENTS, *ro'-dents*—etym., *rodo*, to gnaw—animals, like the rat or squirrel, whose teeth are adapted to gnawing.

ROTIFERA, *ro-tif"-er-a*—etym., *rota* (*ro'-ta*), wheel, and *fero*, to bear—infusorial animals whose cilia move in a rotary manner.

RUBELLITE, *ru'-bel-ite*—etym., *rubrum*, red—a red species of tourmaline.

RUMINANTS, *ru'-min-ants*—etym., *rumino*, to chew the cud—animals which chew the cud.

RUTILE, *ru'-til*—etym., *rutilus* (*ru'-til-us*), red—a reddish ore of titanium.

S.

SAHLITE, *sa'-lite*. See AUGITE, of which it is a variety.

SALAMANDER, *sal'-a-man-der*—etym., *salamandra*, the ancient name of a mythic animal supposed to be capable of resisting fire—the modern popular name of a batrachian resembling the lizard and the frog.

SAPPHIRE, *saf'-fer*—etym., σαπφειρος, *sap-fi'-ros*, supposed to mean smooth, lustrous—pure crystallized alumina, a mineral of a bluish tint.

Sard—etym., Σαρδα, *sard'-a*, Sardinia, or Σαρδις, *sardis*—a variety of chalcedony.

Saurian, *saw'-re-an*—etym., *sauros* (*saw'-ros*), lizard, resembling a lizard.

Scansores, *scan-sō'-reez*—etym., *scando*, to climb—climbers, as the woodpecker.

Scapolite, *scap'-o-lite*—etym., σκαπος, *sca'-pos*, rod, and λιθος, stone—a silicious mineral, occurring in four or eight-sided prisms, terminated by low pyramids.

Selenite, *sel'-en-ite*—etym., σεληνη, *se-lē'-nee*, moon—a laminated, lustrous gypsum, so called from its partial resemblance to the aspect of the moon.

Serpentine, *ser'-pen-tin*—etym., *serpens*, serpent—a magnesian stone, shaded and spotted somewhat like a serpent's skin.

Shale—etym., *schale*, *shale*, shell—a fine-grained slaty rock.

Sigillaria, *sij-il-ā'-re-a*—etym., *sigillum* (*se-jil'-lum*), seal—fossil trees whose bark is covered with impressions as if made with a seal.

Silurian, *sil-ū'-re-an*, the system of rocks first observed in the region of the *Silures* (*sil-ū'-reez*), the ancient inhabitants of the border land of England and Wales.

Siphuncle, *sif'-unk-l*—etym., *siphon*, a conducting pipe—the tube which runs through chambered shells, as in the *nautilus*.

Sodium, *sō-de-um*, the metallic base of soda.

Spar—etym., *spar*, bar or beam—any lustrous earthy mineral which breaks with regular surfaces.

Situ, *si'-tu* ("In situ")—etym., *situs* (*si'-tus*), place—in place, in its original or natural situation, not transported.

Spatangoid, *spat-ang'-goid*—etym., σπαταγγος, *spat-ang'-gos*, pail or vessel—a sea-urchin, so named from its peculiar shape.

SPATHIC, *spath'-ic*—etym., *spath*, slice or plate—in laminæ or plates.

SPECULAR—etym., *speculum* (*spec'-u-lum*), mirror—having a smooth, reflecting surface.

SPHAGNUM—etym., *sphagnum*, bog-moss.

SPORE—etym., σπορος, *spor-os*, sowing—the part of flowerless plants which serves instead of seeds, as the specks on the back of the fern leaf.

STALACTITE, *stal-ac'-tite*—etym., σταλαζω, *stal-a'-zo*, to drop—mineral accumulations formed by the dropping of water, containing lime in solution, from the roof of a cavern.

STALAGMITE, *stal-ag'-mite*, σταλαγμος, *stal-ag'-mos*, dripping, dropping—a deposit of calcareous or other matter, made by water dropping on the floor of a cavern.

STAUROTIDE, *staw'-ro-tide*—etym., σταυρος, *staw'-ros*, cross, and ειδος, form—a silicious mineral, found in crystallized prisms, sometimes intersecting each other at right angles or in the form of a cross.

STEATITE, *ste'-a-tite*—etym., στεαρ, *ste'-ar*, fat—a talcose rock, composed of silica and magnesia. It feels like tallow or soap to the touch, and hence its common name "soapstone."

STILBITE—στιλβω, *stil'-bo*, to shine—a mineral found in amygdaloid in whitish lustrous crystals.

STIGMARIA, *stig-ma'-re-a*—etym., *stigma*, mark or brand—plants of the coal period having curiously-marked stems.

STRIA, *stri'-a*—etym., *stria*, plural *striæ*, scores or grooves—regular rows of scratches on the surface of rocks.

STRIATED, scored, grooved, or furrowed.

STRONTIAN, *stron'-she-an*—etym., *Strontian*, *stron'-she-an*, in Argyleshire, Scotland, where this mineral was first found—a heavy, earthy mineral, of whitish color.

STRONTIUM—etym. as above—the metal which is the base of the above mineral.

STRATA, *stra'-ta*—etym., *stratum* (*strā'-tum*), spread or layer—earthy or rocky matter in layers.
SULPHATE, *sul'-fate*—etym., *sulphur*—a salt formed by sulphuric acid combined with a base, as sulphate of lime.
SULPHURET, *sul-fu'-ret*, a combination of sulphur with a base.
SYENITE, *si'-en-ite*—etym., *Syena* (*si-e'-na*), a region of Egypt, where this rock was early observed.

T.

TABULAR, *tab'-u-lar*—etym., *tabula* (*tab'-u-la*), table—of flat surface.
TALC, *tâlc*—etym., *talc*, tallow—a smooth, lustrous, laminated, magnesian mineral.
TALCOSE, *talc'-ose*—etym., *talc*, tallow—containing talc. See TALC.
TELLURIUM, *tel-u'-re-um*—etym., *Tellus*, earth—a grayish metal, combined with gold and silver in the ores.
TEREBRATULA, *ter-e-brat'-u-la*—etym., *terebro*, *ter'-e-bro*, to bore—a genus of bivalve molluscs with a perforated shell.
TESSELATED, *tes'-sel-a-ted*—etym., *tessela*, a small square stone—checkered or divided into small squares.
TESTACEA, *tes-ta'-she-a*—etym., *testa*, shell—animals with a shelly covering.
TETRABRANCHIATES, *tet-ra-brank'-e-ates*— etym., τετρα, *tet'-ra*, four, and βραγχιον, *brank'-e-on*, gill—cephalopods having four gills.
TETRADECAPODS, *tet-ra-dec'-a-pods*—etym., τετρα, *tet'-ra*, four, δεκα, *deck'-a*, ten, and ποδα, *pod'-a*, feet—so called in allusion to the number of locomotive appendages.
THALLOPHYTES, *thal'-lo-fites*—etym., θαλλος, *thal'-los*, young branch, and φυω, *fu'-o*, to grow—a class of flowerless plants.
TITANIUM, *ti-ta'-ne-um*, a metal of deep blue color.

TOPAZ—etym., *Topazios, to-pā'-ze-os*, the small island in the Arabian Gulf, whence the ancients obtained the precious stone of this name.

TOURMALINE, *toor'-mă-lin*—etym., *Tour'-mă-mal*, the Cingalese name of a mineral, found in three and six-sided prisms of various colors, and much valued for jewellery.

TRACHYTE, *tra'-kite*—etym., τραχυς, *tra'-kuse*, rough—a variety of lava, rough to the touch.

TRAP, etym., *trappe (trap'-pay)*, stair—a rock often presenting the form of blocks resting on each other as the steps of a stair.

TRAPEZOHEDRON, *trap-e-zo-he'-dron*—etym., τραπεζιον, *trap-e'-ze-on*, table, and εδρα, *hed'-ra*, side—a solid bounded by twenty-four equal and similar trapeziums.

TREMATODS, *trem'-a-tods*—etym., τρῆμα, *tre'-ma*, pore—worms having suctorial pores.

TREMOLITE, *trem'-o-lite*, white hornblende.

TRILOBITE, *tri'-lo-bite*—etym., τρεις, *trice*, three, and λοβος, *lob'-os*, lobe—a fossil crustacean, named from its form.

TUFA, *tu'-fa*—etym., *tufo (tu'-fo)*, porous—soft, porous stone; volcanic rock of sandy, earthy, or basaltic material.

TUNICATA, *tu-ne-cā'-ta*—etym., *tunica (tu'-ne-ca)*, coated—having a membranous coating.

V.

VERD-ANTIQUE, *verd-an-teek'*—etym., *viridis (vir'-e-dis)*, green, and *antiquus (an-ti'-qwus)*, ancient—serpentine, also a greenish porphyry, of a tint resembling that of the green incrustation on ancient coins.

VERTEBRÆ, *ver'-te-bre*—etym., *verto*, to turn—the bony joints enclosing the spinal marrow; parts of the backbone.

VERTEBRATES, *ver'-te-brates*—etym., *vertebræ (ver'-te-bree)*, joints of the spinal column or backbone—animals having such a column.

VITREOUS, *vit'-re-ous*—etym., *vitrum*, glass—glassy.

W.

WEALDEN, *weel'-den*—named from the locality in England where it was first observed.

Z.

ZEOLITE, *ze'-o-lite*—etym., ζεω, *ze'-o*, to foam, and λιθος, *li'-thos*, stone—a family of minerals, found in the cavities of amygdaloids, &c., and named from the appearance which they exhibit before the blow-pipe.

ZIRCON—a mineral occurring in square prisms, with pyramidal terminations.

ZOÖLOGY, *zo-ol'-o-gy*—etym., ζωον, *zo-on*, animal, and λογος, *log'-os*, discourse—the science which treats of the animal kingdom.

ZOÖPHYTE, *zo'-o-fite*—etym., ζωον, *zo'-on*, animal, and φυτον, *fu'-ton*, plant—a term applied to some polyps, from their apparently ambiguous appearance, resembling both animals and plants.

INDEX.

INDEX.

Acalephs, 107.
Acephals, 113.
Acrogens, 95.
Actinia, 104.
Actinoids, 107.
Actinolite, 40.
Agate, 32.
Alabaster, 43.
Albite, 34.
Alcyonoids, 107.
Alluvium, 202.
Aluminium, 25.
Amazon river, 225.
Amethyst, 31.
Ammonite, 176.
Amygdaloid, 77.
Andalusite, 46.
Angiosperms, 93.
Animal kingdom, 102.
Anophytes, 97.
Anterior members of vertebrates, 120.
Anthracite, 166.
Anticlinal axis, 86.
Antiquity of the earth, 268.
Apatite, 45.
Aquamarine, 54.
Aqueous agencies, 217.
Arachnids, 119.
Arsenical iron pyrites, 59.
Articulates, 117.
Asbestus, 40.

Asteroids, 111.
Atoll, 262.
Augite, 41.

Barnacles, 118.
Basalt, 75.
Basin, Franconia Notch, 223.
Batrachians, 122.
Bedford Ravine, 222.
Belemnites, 177.
Beryl, 54.
Bimana, 125.
Birds, 123.
Bituminous Coal, 166.
Bloodstone, 33.
Botany, 14.
Boulders on the Jura, 234.
Brachiopods, 114.
Brahmapootra, 226.
Branches or types, 103.
Brontozoum, 173.
Bryozoa, 113.

Calamites, 158.
Calcareous marl, 261.
Calcareous tufa, 37.
Calc spar, 38.
Carbon, 22.
Carbonic acid, 23.
Carboniferous coral, 163.
Carboniferous crinoids, 162.
Carboniferous fish, 164.

(315)

316   INDEX.

Carboniferous molluscs, 163.
Carboniferous system, 155.
Carnelian, 32.
Carnivora, 125.
Cause of volcanic eruption, 251.
Celestine, 44.
Cephalopods, 115.
Cetaceans, 125.
Chalcedony, 32.
Chalk, 37.
Chelonians, 123.
Chemical constitution of the earth, 20.
Chemistry, 14.
Chlorine, 22,
Chondrodite, 50.
Chromate of iron, 59.
Chrysoprase, 32.
Classes, 103.
Classification of animals, 126.
Classification of plants, 95.
Classification of the rock formations, 139.
Clay slate, 82.
Clay-stones, 205.
Club-mosses, 96.
Coal formation, 165.
Coal plants, 156-160.
Composition of coral reefs, 266.
Concluding remarks, 268.
Conformable strata, 86.
Conglomerate, 83.
Copper, 61.
Copper pyrites, 61.
Coral, 104.
Coral island, 262.
Coral reefs, 261.
Coral reefs, regions of, 265.
Coral reefs, composition of, 266.
Coral reefs, rate of growth of, 266.
Corundum, 50.
Cotopaxi, 250.
Cotyledon, 91.
Crabs, 118.
Cretaceous cephalopods, 183.
Cretaceous echinoderms, 182.
Cretaceous system, 181.
Crinoids, 110.
Crinoid from St. Louis, 133.

Crustaceans, 117.
Cryptogamous plants, 95.
Crystallography, 28.
Ctenoids, 122.
Ctenophoræ, 110.
Cursores, 124.
Cuttle-fish, 116.
Cycas, 93.
Cycloids, 122.

Days of creation, 277.
Decapods, 118.
Dendrite, 137.
Density of the earth, 17.
Description of the stratified rocks, 81.
Description of the unstratified rocks, 68.
Development hypothesis, 279.
Devonian acephals, 152.
Devonian corals, 151.
Devonian fishes, 154.
Devonian gasteropods, 122.
Devonian trilobites, 153.
Diamond, 23.
Dibranchiates, 116.
Dicotyledons, 92.
Dikes at Cohasset, Mass., 78.
Dinornis, 207.
Dip, 84.
Discophoræ, 108.
Distribution of animals, 127.
Distribution of plants, 99.
Dolomite, 36.
Drift, 192.
Drift-wood, 230.

Earth as a planet, 16.
Earthquakes, 253.
Echinoderms, 110.
Echinoids, 112.
Edentata, 125.
Elements, 20.
Emery, 51.
Endogens, 94.
Entomostraca, 117.
Epidote, 49.
Equisetaceæ, 96.
Etna, 243.

INDEX.

Exogens, 91.
Extinct horse, 208.

Families, 103.
Faults, 87.
Feldspar, 33.
Ferns, 95.
Fingal's Cave, 76.
Fishes, 121.
Fluorine, 22.
Fluor spar, 45.
Fool's gold, 59.
Foot-marks of Connecticut Valley, 170.
Foraminifera, 115.
Forward members of vertebrates, 120.
Fossils, 132.
Fossil fruit, 186.
Frost, 217.
Fungi, 98.

Galena, 60.
Ganges, 226.
Ganoids, 122.
Garnet, 48.
Gasteropods, 114.
Gasteropods proper, 115.
Genera, 103.
General statement, 13.
Genessee river, 218.
Geodes, 30.
Geological changes, 216.
Geological range of animals, 213.
Geological range of plants, 212.
Geology, 13.
Geysers, 256.
Glacier of Viesch, 231–232.
Glaciers, 228.
Gneiss, 81.
Gold, 63.
Graham Island, 250.
Grallæ, 123.
Granite vein, Williams' Hill, 72.
Granite, 68.
Graphic granite, 68.
Great Plan, 282.
Greenland, 256.
Greenstone, 74.

Gymnosperms, 94.
Gypsum, 42.

Hardness, scale of, 27.
Heavy spar, 44.
Herculaneum, 241.
Heterocercal, 122.
Heteropods, 115.
Holothurioids, 113.
Homocercal, 122.
Hornblende, 40.
Hornblende slate, 82.
Horse-shoe crab, 118.
Hyacinth, 56.
Hydrogen, 21.
Hydroids, 108.

Icebergs, 234.
Ichthyosaurus, 178.
Ideal section, 142.
Idocrase, 48.
Igneous agencies, 239.
Iguanodon, 184.
Indicolite, 53.
Infusoria, 127.
Insectivora, 125.
Insects, 119.
Insessores, 124.
Iolite, 53.
Irish elk, 208.
Iron, 56.
Iron, brown, 58.
Iron, magnetic, 57.
Iron, oxyd of, 57.
Iron pyrites, 59.
Iron, spathic, 58.
Iron, specular, 57.

Jasper, 33.
Joints, 84.
Jupiter Serapis, 256.
Jorullo, 244.
Jura Mountains, 284.

Kaolin, 34.
Kilauea, 246.
Kyanite, 47.

Lagoon, 202.

INDEX.

Laminæ, 84.
Lamellibranchiates, 114.
Lava, 252.
Leaf arrangement, 259.
Lepidodendron, 159.
Lepidolite, 35.
Life periods, 140.
Limestone, 36.
Lobster, 118.

Madrepore, 107.
Magnesium, 25.
Malachite, 61.
Mammals, 124.
Man produces changes, 268.
Marble, 37.
Marsupials, 125.
Mastodon, 209.
Mauna Loa, 246.
Medusæ, 103.
Megatherium, 209.
Metal, a native, 56.
Metals and metallic ores, 56.
Metals and metalloids, 20.
Metamorphic rocks, 67.
Mica, 35.
Mica slate, 81.
Mineral, a simple, 26.
Mineral constitution of the earth, 26.
Mineralogy, 14.
Mississippi river, 225.
Modified drift, 203.
Molar of elephant, 210.
Molar of mastodon, 209.
Molluscs, 113.
Molybdenum, 62.
Monkeys, 128.
Moonstone, 35.
Moraines, Lateral, 230.
Moraines, Medial, 230.
Moraines, Terminal, 232.
Myriopods, 119.

Natatores, 123.
Nature has not repeated herself, 280.
Nautilus, 116.
New red sandstone system, 169.
Niagara, 219.

Nile, 225.
Nitrogen, 22.
Non-metallic elements, 20.
Nummulitic limestones, 189.
Nyöe, 249.

Ocean currents, 227.
Old red sandstone, 150.
Onyx, 32.
Oölitic cephalopods, 177.
Oölitic echinoderms, 176.
Oölitic reptiles, 177.
Oölitic system, 175.
Ophidians, 123.
Orders, 103.
Ore, 56.
Otozoum, 173.
Oxyd of iron, 43, 57.
Oxygen, 21.

Pachyderms, 125.
Paleontology, 132.
Paleozoic period, characterized by, 214.
Peat, 259.
Pele's hair, 252.
Pentacrinus Caput-Medusæ, 111.
Petrifactions, 134.
Phænogamous plants, 91.
Phosphorus, 24.
Placoids, 121.
Plan revealed, 278.
Plesiosaurus, 178.
Plumbago, 23.
Polished quartz, 198.
Polyps, 103.
Pompeii, 241.
Porphyry, 74.
Potassium, 25.
Pot-holes, 223.
Pot-holes in Orange, N. H., 224.
Prase, 31.
Preparation for man, 281.
Pterodactyl, 179.
Pteropods, 115.
Pumice, 252.
Pyroxene, 41.

Quadrumana, 125.

INDEX.

Quartz, 26.
Quartz, ferruginous, 32.
Quartz, granular, 33.
Quartz rock, 82.
Quartz, rose, 31.
Quartz, smoky, 32.

Radiates, 103.
Rains, 217.
Raptores, 124.
Rasores, 124.
Rate of growth of coral reefs, 266.
Reef, fringing, 261.
Reef, barrier, 261.
Relative age of mountains, &c., 214.
Relative age of veins or dikes, 80.
Reptiles, 122.
Rhizodonts, 123.
Rivers, 218.
Rivers transporting the lands to the sea, 225.
Rock crystal, 28.
Rocking stones, 194.
Rock striated by a glacier, 234.
Rocks which compose the earth, 65.
Rodents, 125.
Rotifera, 117.
Rubellite, 53.
Ruby, 50.
Ruminants, 125.
Rutile, 62.

Sabrina, 250.
Sandstone, 83.
Sandwich Islands, 251.
Same strata dipping in opposite directions, 88.
Sapphire, 50.
Sard, 32.
Sardonyx, 32.
Saurians, 123.
Scandinavia, 255.
Scansores, 124.
Scapolite, 46.
Sea-anemone, 103.
Sea-urchins, 112.
Secondary period, characterized by, 214.

Sections of coral reefs and a coral island, 264.
Selenite, 42.
Serpentine, 41.
Shrimp, 118.
Siberian elephant, 207.
Sigillaria, 160.
Silicious marl, 260.
Silicon, 25.
Silurian acephals, 116.
Silurian cephalopods, 147.
Silurian coral, 145.
Silurian crinoids, 145.
Silurian gasteropods, 147.
Silurian system, 143.
Silurian trilobites, 148.
Silver, 63.
Skapter Jokul, 245.
Sodium, 25.
Species, 103.
Spinel, 50.
Sponges, 107.
Springs, 218.
Stalactites, 37.
Stalagmite, 38.
Star-fish, 111.
Staurotide, 47.
Stone book, 173.
Strata folded, 88.
Stratified rocks, 65.
Striated quartz, 197.
Strike, 85.
Sulphur, 23.
Sumbawa, 245.
Sunstone, 35.
Squids, 116.
Syenite, 73.
Synclinal axis, 86.

Table Rock, 219.
Tabular view of stratified and unstratified rocks, 141.
Talc, 41.
Talcose slate, 82.
Temperature of the earth, 18.
Tertiary acephals, 188.
Tertiary and modern, characterized by, 214.
Tertiary corals, 187.

Tertiary fishes, 190.
Tertiary gasteropods, 188.
Tertiary mammals, 191.
Tertiary system, 185.
Terraces, 205.
Tetrabranchiates, 115.
Tetradecapods, 118.
Thallophytes, 98.
Thermal springs, 252.
Tides, 227.
Tin ore, 62.
Topaz, 54.
Tourmaline, 52.
Trachyte, 77.
Trap rocks, 75.
Tree ferns, 97.
Tremolite, 40.
Trenton Falls, 218.
Tunicata, 114.
Turtles, 123.
Types, 103.

Unconformable strata, 86.
Unstratified rocks, 66.

Valley of Switzerland, 234.
Vegetable kingdom, 91.
Veins or dikes, 70.
Verd-antique, 42.
Vertebrates, 119.
Vertical movements without earthquakes, 255.
Vesuvius, 240.
Volcanoes, 239.

Waves, 226.
Worms, 117.

Zeolite family, 51.
Zinc, 60.
Zircon, 56.
Zoölogy, 14.
Zoöphytes, 104.

THE END.

CATALOGUE

OF

Approved School and College Text-Books.

PUBLISHED BY E. H. BUTLER & CO.,

137 South Fourth St., Philadelphia.

Goodrich's Pictorial History of the United States. A Pictorial History of the United States, with notices of other portions of America. By S. G. GOODRICH, author of "Peter Parley's Tales." For the use of Schools. Revised and improved edition, brought down to the present time (1860). Re-written and newly illustrated. 1 vol. 12mo., embossed backs. Upwards of 450 pages. Price $1.18

Goodrich's American Child's Pictorial History of the United States. An introduction to the author's "Pictorial History of the United States." Will be published in July, 1860.

Goodrich's Pictorial History of England. A Pictorial History of England. By S. G. GOODRICH, author of "Pictorial History of the United States," etc. Price $0.94

Published by E. H. BUTLER & CO., Philadelphia.

Goodrich's Pictorial History of Rome. A

Pictorial History of Ancient Rome, with sketches of the History of Modern Italy. By S. G. GOODRICH, author of "Pictorial History of the United States." For the use of Schools. Revised and improved edition. . Price $0.94

Goodrich's Pictorial History of Greece. A

Pictorial History of Greece; Ancient and Modern. By S. G. GOODRICH, author of "Pictorial History of the United States." For the use of Schools. Revised edition. . Price $0.94

Goodrich's Pictorial History of France. A

Pictorial History of France. For the use of Schools. By S. G. GOODRICH, author of "Pictorial History of the United States." Revised and improved edition, brought down to the present time. . . . . . Price $0.94

Goodrich's Parley's Common School History

of the World. A Pictorial History of the World; Ancient and Modern. For the use of Schools. By S. G. GOODRICH, author of "Pictorial History of the United States," etc. Illustrated by engravings. . Price $1.13

Goodrich's First History. The First History.

An Introduction to Parley's Common School History. Designed for beginners at Home and School. Illustrated by Maps and Engravings. By S. G. GOODRICH, author of the Pictorial Series of Histories, etc. . . . Price $0.38

Published by E. H. BUTLER & CO., Philadelphia.

Hows' Ladies' Reader. The Ladies' Reader.
Designed for the use of Ladies' Schools and Family Reading Circles; comprising choice selections from standard authors, in Prose and Poetry, with the essential Rules of Elocution, simplified and arranged for strictly practical use. By JOHN W. S. Hows, Professor of Elocution. . . Price $1.13

Coppee's Elements of Logic. Elements of
Logic. Designed as a Manual of Instruction. By HENRY COPPEE, A. M., Professor of English Literature in the University of Pennsylvania; and late Principal Assistant Professor of Ethics and English Studies in the United States Military Academy at West Point. . . . . Price $0.75

Coppee's Elements of Rhetoric. Elements
of Rhetoric. Designed as a Manual of Instruction. By HENRY COPPEE, A. M., author of "Elements of Logic,".etc. New edition, revised. . . . . . Price $1.00

Tenney's Geology. Geology; for Teachers,
Classes, and Private Students. By SANBORN TENNEY, A. M., Lecturer on Physical Geography and Natural History in the Massachusetts Teachers' Institutes. Illustrated with Two Hundred Wood Engravings. . . Price $1.13

Stockhardt's Chemistry. The Principles
of Chemistry, illustrated by Simple Experiments. By Dr. JULIUS ADOLPH STÖCKHARDT, Professor in the Royal Academy of Agriculture at Tharand, and Royal Inspector of Medicine in Saxony. Translated by C. H. PEIRCE, M. D. Fifteenth Thousand. . . . . . . Price $1.96

Published by E. H. BUTLER & CO., Philadelphia.

Reid's Essays on the Intellectual Powers of

Man. Essays on the Intellectual Powers of Man. By THOMAS REID, D. D., F.R.S.E. Abridged, with notes and illustrations from Sir WILLIAM HAMILTON and others. Edited by JAMES WALKER, D. D., President of Harvard College.   Price $1.31

Stewart's Philosophy of the Active and

Moral Powers of Man. The Philosophy of the Active and Moral Powers of Man. By DUGALD STEWART, F.R.SS. Lond. and Ed. Revised, with omissions and additions, by JAMES WALKER, D. D., President of Harvard College.   Price $1.31

Mitchell's First Lessons in Geography.

First Lessons in Geography; for young children. Designed as an Introduction to the author's Primary Geography. By S. AUGUSTUS MITCHELL, author of a Series of Geographical Works. Illustrated with maps and numerous engravings.   Price $0.38

Mitchell's Primary Geography. An Easy

Introduction to the study of Geography. Designed for the instruction of children in Schools and Families. Illustrated by nearly one hundred engravings and sixteen colored maps. By S. AUGUSTUS MITCHELL. . . . . Price $0.42

Mitchell's New Intermediate Geography.

An entirely new work. The maps are all engraved on copper, in the best manner, and brought down to the present date. It is profusely illustrated with beautiful engravings, and is the most complete quarto Geography ever issued in the world. . . . . Price $1.12½

Published by E. H. BUTLER & CO., Philadelphia.

www.ingramcontent.com/pod-product-compliance
Lightning Source LLC
Chambersburg PA
CBHW030745230426
43667CB00007B/854